The Extractive Industries Sector

A WORLD BANK STUDY

The Extractive Industries Sector

Essentials for Economists, Public Finance Professionals, and Policy Makers

Håvard Halland, Martin Lokanc, and Arvind Nair, with
Sridar Padmanabhan Kannan

WORLD BANK GROUP

ISBN (paper): 978-1-4648-0492-2
ISBN (electronic): 978-1-4648-0493-9
DOI: 10.1596/978-1-4648-0492-2

Cover art: © Cor Laffra, www.corlaffra.com. Used with permission. Further permission required for re-use.
Cover design: Debra Naylor, Naylor Design, Inc.

Library of Congress Cataloging-in-Publication Data has been requested

Contents

Figures

Tables

Acknowledgments

This first of the two-volume study *Essentials for Economists, Public Finance Professionals, and Policy Makers* was prepared by Håvard Halland, Martin Lokanc, and Arvind Nair, with contributions from Sridar Padmanabhan Kannan, all of the World Bank. Its production was led by Håvard Halland. The volume is a joint product of the World Bank's Governance Global Practice and the Energy & Extractives Global Practice. It draws on a large number of World Bank and publicly available documents. Particularly relevant World Bank documents, or documents for which the World Bank holds the copyright, are in certain cases summarized or condensed. Where a single World Bank copyrighted source is available, the study extracts and summarizes relevant material from this source. In chapter 5, the sections on geodata and cadastre are in this way based on BGS International (2012) and Ortega Girones, Pugachevsky, and Walser (2009), respectively, and the section on competitiveness reproduces material from Gammon (2007). Also in chapter 5, the sections on ownership and on extractive industries (EI) taxation summarize and condense material from the *EI Source Book* (Cameron and Stanley 2012). Material that draws directly on non–World Bank sources is otherwise reproduced in boxes, with reference to the original publication. The objective of this volume is not to present original research, but rather to survey and summarize insights from an extensive body of literature, and condense these insights into an easily readable format. For the interested reader, references to the most relevant original documents are sometimes provided at the start of the appropriate sections.

The authors are grateful to the Governance Partnership Facility and its donor partners—the U.K. Department for International Development (DFID), the Australian Department of Foreign Affairs and Trade (DFAT), the Netherlands' Ministry of Foreign Affairs, and Norway's Ministry of Foreign Affairs—for providing full funding for this work. The authors are also grateful to Yue Man Lee for the formal peer review of an earlier draft and invaluable input that significantly improved the final product. Bryan Land, Marijn Verhoeven, Nicola Smithers, and Adrian Fozzard also provided extremely useful feedback on earlier drafts, and the authors greatly appreciate managerial support from Robert Beschel and Michael Jarvis. Michael Stanley, Boubacar Bocoum, Adriana Eftimie, Remi Pelon, and Noora Arfaa shared numerous highly relevant source documents, and provided crucial insight and advice. Editorial work by Fayre Makeig made the study much easier to read. Xin Tong's design added greatly to its visual appeal. Copyrights for tables and figures drawn from other sources were obtained by Catherine Lips.

The authors are grateful to the companies, organizations, and individuals who generously granted usage rights for copyrighted figures and tables, including Anglo American plc; William Ascher; Barrick Gold Corporation; Committee for Mineral Reserves International Reporting Standards; European Union; J.P. Morgan Commodities Research; Paul Jourdan; Metal Bulletin Research; MinEx Consulting Pty Ltd.; Ministry of Mines and Petroleum, Government of Afghanistan; National Academies Press, National Academies of Sciences; Natural Resources Canada; Organisation for Economic Co-operation and Development; Princeton University Press; Revenue Watch Institute, now the Natural Resource Governance Institute; Rio Tinto; Society for Mining, Metallurgy & Exploration; Society of Petroleum Engineers; Taylor & Francis Books UK; United Nations Economic Commission for Europe; and Wood Mackenzie.

All opinions, errors, and omissions are the authors' own.

About the Authors

Håvard Halland is a natural resource economist at the World Bank. His research and advisory work focus on the economics and finance of the extractive industries sector. Research and policy agendas include resource-backed infrastructure finance, sovereign wealth fund policy, extractive industries revenue management, and fiscal management in resource-rich countries. He is an author or joint author of academic and policy research papers, book chapters, magazine articles and blogs, and regularly presents at international conferences and seminars. Prior to joining the World Bank, he was a delegate and program manager for the International Committee of the Red Cross in the Democratic Republic of the Congo and Colombia. He earned a PhD in economics from the University of Cambridge.

Martin Lokanc is a mining specialist in the World Bank's Energy and Extractives Global Practice. He currently supports and leads mining sector development activities in Botswana, Zimbabwe, Zambia, Romania, Afghanistan, Malawi, and Bhutan, in addition to managing a number of research and global knowledge projects. Trained as a mining engineer and economist, he has global experience in mine development, mining strategy, mineral and energy economics, mining finance, and economic development, gained over more than 15 years working in the private sector and with the World Bank Group. Martin holds a BSc in mining engineering from the University of Alberta, Canada; an MSc in mining engineering, specializing in mineral economics, from the University of the Witwatersrand, South Africa; and is currently completing a PhD in mineral and energy economics from the Colorado School of Mines in the United States.

Arvind Nair is an economist and consultant to the World Bank, based in Indonesia, where he focuses on natural resource sector engagement and specifically on revenue collection, the Extractive Industries Transparency Initiative, and the macroeconomic impacts of the extractive sector, as part of the Macroeconomics and Fiscal Management Global Practice. Prior to joining the World Bank, he served as an Overseas Development Institute Fellow in the budget office of Sierra Leone's Ministry of Finance and as a research associate with the Institute for Financial Management and Research in India. Arvind holds a master's in public administration and international development from the Harvard Kennedy

School of Government; an MSc in economics for development from Oxford University; and a BA, with honors, in mathematics from Swarthmore College.

Sridar P. Kannan is an operations analyst and consultant to the World Bank, where he helps design and implement projects and creates knowledge material on behalf of the Energy & Extractives Global Practice. Prior to joining the World Bank, he worked as a corporate and infrastructure finance counsel in the Tata Group of Companies, Mumbai, India. Sridar holds an LLM in international business and economic law from Georgetown University; a BA and LLB, with honors, from National Law University, Jodhpur, India; and a certificate in World Trade Organization studies from Georgetown University's Institute of International Economic Law.

Abbreviations

AfDB	African Development Bank
AGS	Afghan Geological Survey
CDA	community development agreement
CES	constant elasticity of substitution
CGS	Council for Geoscience
CIM	Canadian Institute of Mining, Metallurgy and Petroleum
CIMVal	Canadian Institution of Mining Valuation
CMMI	Council of Mining and Metallurgical Institutes
COG	cutoff grade
CRIRSCO	Committee for Mineral Reserves International Reporting Standards
Cu	copper
DRC	Democratic Republic of the Congo
EI	extractive industries
EIA	environmental impact assessment
EI-TAF	Extractive Industries Technical Advisory Facility (of the World Bank)
EITI	Extractive Industries Transparency Initiative
EMP	environmental management plan
ESIA	environmental and social impact assessment
ESMP	environmental and social management plan
FOB	free on board
FTFs	foundations, trusts, and funds
GDP	gross domestic product
GRA	Ghanaian Revenue Authority
HR	human resources
HRD	human resource development
ICMM	International Council on Mining and Metals
IFC	International Finance Corporation
IMC	Inter-Ministerial Committee

IMF	International Monetary Fund
IoU	intensity of use
IT	information technology
JORC	Joint Ore Reserves Committee
KCM	Konkola Copper Mines
km	kilometer
km^2	square kilometer
LC	letter of credit
LED	local economic development
LTU	large taxpayer unit
M&E	monitoring and evaluation
MBR	Metal Bulletin Research
MDAs	ministries, departments, and agencies
MPIGM	Mongolian Professional Institute of Geoscience and Mining
MTEF	medium-term expenditure framework
NGO	nongovernmental organization
NOC	national oil company
NPV	net present value
NRC	national resource company
O&M	operation and maintenance
OECD	Organisation for Economic Co-operation and Development
OPEC	Organization of the Petroleum Exporting Countries
PERC	Pan-European Reserves and Resources Reporting Committee
PFM	public financial management
PIM	public investment management
PPP	public–private partnership
PSA	production-sharing agreement
PV	photovoltaic
PVC	polyvinyl chloride
R&D	research and development
RFI	resource-financed infrastructure
SAMREC	South African Code for Reporting of Exploration Results, Mineral Resources, and Reserves
SAMVAL	South African Code for the Reporting of Mineral Asset Valuation
SEA	strategic environmental assessment
SEEA	System of Environmental-Economic Accounting
SME	Society for Mining, Metallurgy & Exploration
SMEs	small and medium enterprises

SOE	state-owned enterprise
SPE-PRMS	Society of Petroleum Engineers–Petroleum Resources Management System
STEM	science, technology, engineering, and math
SWF	sovereign wealth fund
UNFC	United Nations Framework Classification for Fossil Energy and Mineral Reserves and Resources
USGS	United States Geological Survey
VA	value addition
VAT	value added tax
WPC	World Petroleum Council

Overview

What Should We Know about the Extractive Industries Sector?

Economists, public finance professionals, and policy makers working in resource-rich countries are frequently confronted with issues that require an in-depth understanding of the extractive industries (EI) sector, its economics, governance, and policy challenges, as well as the implications of natural resource wealth for fiscal and public financial management (PFM). The objective of the two-volume *Essentials for Economists, Public Finance Professionals, and Policy Makers*, published in the World Bank Studies series, is to provide a concise overview of the extractive-related topics that economists, public finance professionals, and policy makers are likely to encounter. Volume I, *The Extractive Industries Sector*, provides an introduction to the sector, including an overview of issues core to its economics, institutional framework, project and investment cycles, and contract management, and a description of the components of sector governance and policy. Volume II, *Fiscal Management in Resource-Rich Countries*, addresses the fiscal challenges typically encountered when managing large revenue flows from the EI sector. Since oil and mineral taxation, including subnational revenue sharing, has been extensively addressed elsewhere, the *Essentials* provide only brief treatment of this topic, while referring the reader to relevant sources.

This initial overview provides a common introduction to the two volumes. To this end, it first outlines several key characteristics and challenges that distinguish the EI sector from other sectors. It then reviews experiences of countries that have undertaken successful extractive-led development, and synthesizes key findings from literature on the so-called resource curse hypothesis (which argues that countries rich in oil and minerals have lower growth and worse development outcomes than their peers). It concludes by introducing the two volumes in turn.

How Does the EI Sector Differ from Other Sectors?

The EI sector occupies an outsize space in the economies of many resource-rich countries. Specifically, it accounts for at least 20 percent of total exports, and at least 20 percent of government revenue, in 29 low-income and lower-middle-income countries. In eight such countries the EI sector accounts for more than 90 percent of total exports and 60 percent of total government revenue (IMF 2012). Meanwhile, the expansion of the extractive sector has spurred investment in these countries, reflected in the quintupling of foreign direct investment in Africa between 2000 and 2012—from $10 billion to $50 billion (UNCTAD 2013).

In principle, the extractive sector is not necessarily more complex than other economic sectors. Companies make holes in the ground from which they extract oil, gas, or minerals to be transported to a processing facility in-country or to an export point. Conveniently, the extracted commodities can be weighed and their quality measured, prices of common commodities are quoted on international exchanges, and the industry is dominated by a tiny number of very large companies (Calder 2014). Nevertheless, the economic, societal, and environmental implications of EI operations pose significant and diverse challenges.

For companies, the exploration and extraction of oil, gas, and minerals involve high levels of geological uncertainty, large initial capital investments, and long exploration and project development periods. The high volatility of oil and mineral prices and the unpredictability of costs, meanwhile, generate price and cost risks. EI projects may also generate high risks to the natural environment. The costs of decommissioning projects and, in some cases, the cleanup of contaminated soil or water, can constitute a significant part of total project costs, and companies will typically be required to post collateral to ensure that funding is available to responsibly decommission the project at the end of its operative life. If not taken into account during the licensing of extraction rights, environmental costs could end up as government liabilities instead of on the company balance sheet. Local-level considerations also include the socioeconomic circumstances and health of populations living in the vicinity of the extractive project. To mitigate potentially adverse social and environmental impacts, and ensure that a share of benefits accrues to affected populations, resource companies may be required to meet specific commitments through community development agreements and community foundations, trusts, and funds.

For governments, the exhaustible, nonrenewable character of oil, gas, and mineral resources poses challenges relevant to the determination of optimal extraction rates; the design of the fiscal regime; and the allocation of resource revenues to investment, consumption, and foreign savings. The exhaustibility of subsoil resources also raises complex questions around intergenerational equity and long-term fiscal sustainability. Fiscal planning is likely to be significantly affected by the time profile of extraction and by expected and actual commodity prices.

In the EI sector, specialized technology and high capital requirements generate barriers to entry. As a result, the sector is dominated by large multinational firms with vertically integrated value chains and specialized intellectual property. In low-income countries, this usually means that high-value machines and equipment for operations are imported, whereas the natural resources they extract are exported. The complexity of large-scale multinational operations requires resource-rich countries to develop adequate institutional capacity to establish and operate efficient contracting, legal, and fiscal regimes and to oversee company operations. At the other end of the spectrum, small-scale artisanal mining may provide livelihoods for low-income families, but extensive use of toxic chemicals could result in large liabilities for the government if it must pick up the tab for cleanup.

The locations of natural resource extraction sites are predetermined by geography; extraction projects (unlike manufacturing, for example) cannot be shifted to less costly locations. The global production value chain, meanwhile, involves complex organizational and financing structures that may take advantage of tax treaties and innovative financing mechanisms to ensure that transactions are tax efficient. From the perspective of public revenue management, the global value chain implies challenges related to transfer pricing and beneficial ownership.[1]

The extractive sector is characterized by exceptional profits—and substantial rents, defined as the difference between production costs (including "normal" profits) and revenue from sales. The rents can be highly volatile, as they respond to fluctuations in commodity prices and extraction costs, presenting further challenges to the design of fiscal regimes. Resource prices not only fluctuate to extremes, but do so unpredictably. The fact that countries' resource revenues are typically generated by exports, in the form of foreign currency inflows, puts pressure on exchange rates, with potentially significant effects on competitiveness and macroeconomic stability.

The EI sector, more than many others, depends for its efficient functioning on a complex ecosystem of governmental institutions and functions. The establishment of a fertile EI investment climate requires not only good and well-implemented legal and regulatory regimes but also a functional geodata information base and a mineral rights cadastre. The multifaceted character of the sector is reflected by the involvement of a large number of ministries and public entities whose coordination may be highly complex. Efficient extractive-based economic development requires the effective cooperation of these public entities while drawing on the specialized capacity of each. Yet, cooperation often suffers as individual entities seek to maintain control of their share of the extractive portfolio—and revenues.

Although there is no single explanation for the resource curse, many elements of successful natural-resource-based growth are by now relatively well understood. Countries that have benefited from the EI sector tend to have embraced policies with a common set of characteristics: efficient fiscal regimes and macroeconomic stabilization; the conscientious development of specialized public management capacity in the oil, gas, and mining sectors; and productive investments in infrastructure, human development, and economic diversification. These countries' sustainable and equitable long-term growth has resulted from investing resource revenues in durable assets, and from coordinating diverse economic sectors toward the common goal of resource-based growth. Hence, to optimize the monetary and nonmonetary benefits of oil, gas, and mineral extraction, EI sector policies need to go beyond individual projects, to consider and address the complex set of capabilities needed to ensure the sector's efficient operation and its delivery of optimal benefits to both citizens and the government.

The Blessing, and Curse, of Resource Abundance

Some resource-rich countries have succeeded in converting resource wealth into long-term and equitable economic development, while many others have not. Natural resources have played a fundamental role in the growth of several industrialized economies, including the United Kingdom and Germany, where coal and iron ore deposits were a precondition for the Industrial Revolution. The United States was the world's leading mineral economy from the mid-nineteenth to the mid-twentieth century and in the same period became the world's leader in manufacturing (van der Ploeg 2011). More recently, countries such as Botswana, Chile, and Norway have used abundant oil and mineral resources as the foundation for economic growth. However, in many other countries resource extraction appears to have undermined governance, fed corruption and capital flight, and increased inequality.

Why do some countries succeed in leveraging their natural resources, while others have low growth performance in spite of immense subsoil wealth? This question has been the subject of extensive debate. Sachs and Warner (1995) confirmed a negative relationship between the extractive export share of gross domestic product (GDP) and economic growth. They concluded that resource abundance is associated with slower growth, the relationship that was later labeled the "resource curse." Other authors, using different methods, have disputed the existence of a universal resource curse (Alexeev and Conrad 2009; Brunnschweiler and Bulte 2006; Davis and Tilton 2005). While the existence of such a curse is certainly disputable, it is nevertheless clear that a number of resource-rich developing countries, in spite of growth spikes during periods of particularly high oil and mineral prices, have not been able to translate resource wealth into sustainable long-term growth. As Davis and Tilton (2005) put it:

> While [the question of] whether or not mining usually promotes economic development remains unresolved, there is widespread agreement that rich mineral deposits provide developing countries with opportunities, which in some instances have been used wisely to promote development, and in other instances have been misused, hurting development. The consensus on this issue is important, for it means that one uniform policy toward all mining in the developing world is not desirable…. The appropriate public policy question is not should we or should we not promote mining in the developing countries, but rather where should we encourage it and how can we ensure that it contributes as much as possible to economic development and poverty alleviation.

Although a full review of the literature on the resource curse is beyond the scope of this work, a summary of its main arguments provides useful background. Much of the literature subsequent to Sachs and Warner (1995) concentrates on identifying the mechanisms by which natural resources affect growth.

The relative importance of such mechanisms has been much debated and remains the subject of substantive disagreement.

The culprits most often blamed for the resource curse include "Dutch disease," low or inefficient investment (including in human capital), fiscal indiscipline and high consumption, the decay of institutions, and output volatility generated by the volatility of oil and mineral prices. So-called Dutch disease is often cited. The name alludes to the appreciation of the Dutch currency following oil production in the North Sea in the 1960s and refers to the dynamics by which high production in the extractive sector generates increased demand in the nontradable (services) sector and thus causes currency to appreciate. This appreciation in turn leads to reduced exports from the nonextractive tradable sector (Corden and Neary 1982), which may negatively affect growth.

Institutional quality, as reflected in the rule of law and in the quality of public sector management, is frequently referred to as a possible cause of the resource curse. Political economists point out that in many countries that have found it difficult to generate resource-based growth, the discovery of oil, gas, or minerals was preceded by a legacy of poor governance and weak institutions. Weak institutions offer few checks on rent seeking and corruption. While a small elite may become extremely rich off resource rents, the population as a whole receives few benefits. Mehlum, Moene, and Torvik (2006), for example, distinguish between institutional contexts that are "grabber friendly" and those that are "producer friendly." "Grabbers" of resource revenues are more likely to have free rein where institutions are weak. If serving in a government position is seen as a way to get rich quick, instances of "grabbing" may accelerate. Political economists point out that where incumbent politicians fear removal from office, the administration is likely to extract faster than the socially optimal rate and will borrow against future resource revenues. Common phenomena in such contexts may include capital flight, high private consumption among those in power, and high rates of public spending to benefit favored clients (van der Ploeg 2011). Incumbents who fear losing office may also avoid accumulating public savings—for example, in a sovereign wealth fund (SWF)—that could be raided by a future government, preferring instead to overinvest in partisan projects that increase their own hold on power.

While institutional strength may determine the success of extractive-based development, large revenue flows from the EI sector may degrade institutions. Where large revenue flows occur amid insecure property rights, poorly functioning legal systems, and imperfect markets, they are likely to prompt rent seeking (Torvik 2002). Resource revenues increase the value of being in power. Where they provide funding for autocratic regimes, they can in effect prevent the redistribution of political power toward the middle class, thereby impeding the adoption of growth-promoting policies (Bourguignon and Verdier 2000). In the same vein, the availability of such revenues may encourage elites to block technological and institutional improvements that could weaken their hold on power (Acemoglu and Robinson 2006). In the extreme, disputes over access to natural resources may spark armed conflict. Collier and Hoeffler (2004) estimate that a country whose natural resources compose more than 25 percent of GDP faces a 23 percent probability of civil conflict—against 0.5 percent for a country with no resources.

It can be argued that the political economy literature and its application of economic modelling to the resource sector has substantial weaknesses. Weaknesses of the existing theory include the following:

- The property right over the resource is assumed to be held by the state, and rents flow from the ground without need of investment or effort.
- Despite the models having a purported focus on subsoil resources, they all ignore the finite nature of these resources. No consideration is given to stock constraints, and many models simply assume an infinite resource, produced without any effort or costs. None of the approaches explicitly model a mineral, gas, or oil resource.
- Although state ownership of natural resources is a feature observed in many countries, the models fail to examine the sensitivity of outcomes to different property rights arrangements (for example, private ownership with taxation, some direct state participation, and indigenization policies).

Resource abundance may exacerbate fiscal indiscipline. Sudden revenue windfalls from extractives tend to generate expectations of increased public expenditure, which may prompt the excessive loosening of fiscal policy, and low savings. The result can be public investment in unnecessary or unproductive projects, and increased sovereign debt. In fact, whereas observed and optimal savings rates seem to differ little in nonresource economies, they differ sharply in resource-rich countries (van der Ploeg 2011). Bleaney and Halland (forthcoming, 2015) do not find evidence that natural resource wealth in general promotes fiscal indiscipline. In fact, their results indicate that fuel exporters tend to have a better general government fiscal balance. However, some of the resource-rich countries in their sample have, after oil or mineral discoveries, exhibited severe fiscal indiscipline that cannot be explained by the authors' econometric model. Bawumia and Halland's (forthcoming, 2015) findings testify to the importance of early management of expectations, real fiscal discipline as opposed to a reliance on fiscal rules, full and real (as opposed to nominal) independence of the central bank, as well as the establishment of means to isolate from political pressures the sovereign wealth fund and the government entity responsible for oil revenue projection.

If the volatility of commodity prices—and in turn of public revenue flows from extractives—is passed on to public expenditures and output, it may have a damaging effect on growth. Van der Ploeg and Poelhekke (2010) find that natural resource wealth has a positive direct effect on growth, which is more or less canceled out by the indirect effect of output volatility. In this line of argument, the resource curse would arise from the high volatility of commodity prices, with an effect on growth via output volatility that may be mitigated by financial sector development and openness to trade. Bleaney and Halland (2014) find that the volatility of public expenditure—alongside overall institutional quality—explains slower growth, indicating that resource-rich countries that are able to smooth public expenditures do better than their peers.

Unless there is technology transfer from the EI sector to national industries, resource wealth could contribute to deindustrialization. Yet, some resource-rich countries have achieved broad industrial development even as their currency has appreciated amid large resource exports. Dutch disease thus fails to fully explain the different industrial development trajectories observed across resource-rich countries. Some studies (Gylfason, Herbertsson, and Zoega 1999; Matsuyama 1992) suggest that resource-based industrialization and growth take place if the extractive sector is a source of technology transfer and "learning by doing." Torvik (2001) points to Norway as an example: here, according to his argument, natural resource extraction prompted learning by doing in both the traded and non-traded sectors.

For more complete surveys of the resource curse literature, interested readers are referred to van der Ploeg (2011) and Frankel (2010).

Content of the Two Volumes
The first volume, *The Extractive Industries Sector*, provides an overview of issues core to EI economics; discusses key components of the sector's governance, policy, and institutional frameworks; and identifies the public sector's EI-related financing obligations. Its discussion of EI economics covers the valuation of subsoil assets, the economic interpretation of ore, and the structure of energy and mineral markets. The volume maps the responsibilities of relevant government entities and outlines the characteristics of the EI sector's legal and regulatory frameworks. Specific key functions of the sector are briefly discussed, such as the administration of geodata and cadastre, the characteristics and administration of an efficient EI fiscal regime, contract management and monitoring, and typical requirements for a fertile EI business environment.

The volume also describes the economic and financial structures that underpin environmental and social safeguards, such as the use of financial sureties for decommissioning, and of community foundations, trusts, and funds. The investment of public revenues generated from oil, gas, or minerals is briefly addressed, with a focus on infrastructure, and there is a short discussion of extractive-based economic diversification and local content development. For the interested reader, more specialized publications targeting individual subject areas are sometimes referred to in the first paragraph of relevant chapters and subsections. The interested reader will also find additional material in the appendixes, including on revenue collection, revenue projection, and the management of contingent liabilities, as well as material on resource classification systems, reserve reporting standards, types of economic rents characteristic of the EI sector, the relationship between fiscal policy and economic reserves, and the impact of income changes on commodity demand.

Volume II, *Fiscal Management in Resource-Rich Countries*, addresses critical fiscal challenges typically associated with large revenue flows from the EI sector. The volume discusses fiscal policy across four related dimensions: short-run stabilization, the management of fiscal risks and vulnerabilities, the promotion of long-term sustainability, and the importance of good public financial management and public investment management systems. The volume subsequently examines several institutional mechanisms used to aid fiscal management, including medium-term expenditure frameworks, resource funds, fiscal rules, and fiscal councils. The volume also discusses the earmarking of revenue, resource revenue projections as applied to the government budget, and fiscal transparency, and outlines several fiscal indicators used to assess the fiscal stance of resource-rich countries.[2]

Given the diversity of experiences in resource-rich countries, the topics discussed in the two volumes will be more relevant to some countries than others. Each volume can be read independent of the other, though they address common themes. It is hoped that the information they provide will prove a sound basis for economists, public finance professionals, and policy makers wishing to strengthen the management of the EI sector, and associated fiscal and PFM systems, in their countries.

Notes

1. The IMF's 2014 draft update of the Resource Revenue Management pillar (Pillar IV) of the Fiscal Transparency Code defines a beneficial owner as "the legal entity, or if applicable, the natural person which owns the ultimate economic interest in the holder of a natural resource right within a country, usually through a chain of related parties which may be held in different jurisdictions" (http://www.imf.org/external/np/exr/consult/2014/ftc/pdf/121814.pdf). In the context of tax evasion, corporations may hide their beneficial ownership of one or more related companies so as to avoid scrutiny of alleged arm's-length flows of goods and services between the related companies, or subsidiaries. A beneficial owner of a natural resource would be the legal entity or natural person that owns the holder of extraction rights in a country, potentially via related parties located in different jurisdictions.

2. Subsequent work on public financial management, as relevant to natural resources, is in progress.

Organization of this Volume

This volume describes key components of the extractive industries (EI) sector policy and identifies EI-related financing obligations of the public sector. Chapter 1 lays out the broad objectives of resource-based development. Chapter 2 outlines several fundamentals of the EI sector, including accounting for physical stocks, evaluating subsoil assets, and understanding the structure of commodity markets. Chapter 3 describes the institutional framework that frames the EI sector and the responsibilities of various government entities to ensure resource-based growth. Chapter 4 discusses EI sector investment and project cycles. Chapter 5 describes the main operational functions of EI sector governance: the legal and regulatory framework, the administration of geodata and cadastre, the characteristics and administration of transparent and efficient resource taxation, and the EI business environment. Chapter 6 focuses on sector monitoring and contract management, including fiscal regimes and environmental and social issues. Chapter 7 discusses the investment of public funds arising from resource revenues, and Chapter 8 provides a very brief introduction to extractive-based economic diversification and local content development.

CHAPTER 1

Defining Sector Policy Objectives

The Extractive Industries Value Chain

The nonrenewable and finite character of oil, gas, and mineral resources is the primary premise of an extractive industries (EI) sector policy. Other key characteristics are the sector's high capital intensity, long-lived assets, price-taking producers (prices are set in global markets, and only to a limited extent influenced by individual producers), and geographically immobile investments—as well as the international nature of commodity markets and EI investors. To form the basis for long-term sustainable development, subsoil resources (natural capital) must be invested in, or converted to, more productive forms of capital, while respecting the country's environmental and societal foundations. *Adjusted net savings* measure the true level of savings in a country, taking into account the depletion of subsoil assets such as oil and minerals (as well as other natural capital) and investment in human capital, infrastructure, and other produced capital. Sustained negative adjusted net savings will lead to a reduction in total wealth and in the welfare of the population. Many resource-rich countries, particularly in Sub-Saharan Africa, display very low or negative adjusted net savings (Ross, Kaiser, and Mazaheri 2011).

Both the generation and collection of resource revenues—and, later, their productive investment—require a set of policies that span sectors as well as institutional and human capabilities. The EI value chain (figure 1.1) provides a framework for the governance of the EI sector. It encompasses the award of contracts and licenses, monitoring of operations, enforcement of environmental protection and social mitigation requirements, collection of taxes and royalties, distribution of revenue in a sound manner, and implementation of sustainable development policies and projects. The framework is meant as a tool to support countries in their efforts to translate mineral and hydrocarbon wealth into sustainable development (Mayorga Alba 2009).

Figure 1.1   The Extractive Industries Value Chain: A Framework for Governance

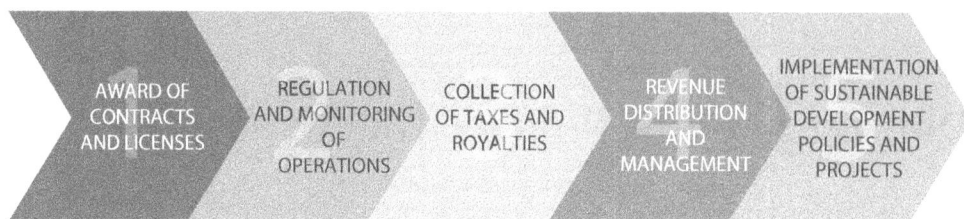

Source: Mayorga Alba 2009.

Improving Revenue Mobilization

For optimal revenue mobilization, many resource-rich developing countries find they need to invest significantly in strengthening their contract negotiation and management capacity, upgrading their resource tax administration, and improving their EI investment climate. As outlined in chapter 6 on monitoring and enforcing contracts, inadequate capacity for contract management and tax administration may lead to losses of revenue if extracted quantities, grades, or prices have been underreported by resource companies. But mobilizing revenue is not limited to contracts, taxes, and royalties; it also involves attracting investment. To attract investment in the EI sector—and thus maximize revenues—a sector policy needs to establish the following:

- The necessary technical capacity to undertake government geological surveys, attract investments in promising areas, and provide information on licensing processes and land-use planning
- Adequate capacity to manage mineral and petroleum exploration and extraction rights, based on a well-functioning and transparent cadastral system
- An attractive business environment (that includes clear, transparent, stable, and predictable mining and/or petroleum legislation) with low barriers to entry, while respecting standards for technical and financial capacity of companies
- Sufficient capacity to negotiate contracts, and to achieve optimal terms for exploration and extraction, where relevant[1]
- Sufficient technical capacity, particularly in mineralogy, to credibly monitor the quantity and quality of mineral production and exports
- Adequate technical capacity in taxation and accounting to credibly manage the collection of taxes, royalties, and fees
- Stability, predictability, transparency, and accountability in contract and revenue management, and of contractual and fiscal terms

In recent years, the issue of capacity for mobilizing resource revenues has been addressed extensively by the International Monetary Fund's (IMF's) Managing

Natural Resource Wealth Topical Trust Fund, as well as by various international nongovernmental organizations (NGOs). The topic of resource taxation is extensively treated in *The Taxation of Petroleum and Minerals* (Daniel, Keen, and McPherson 2010). Financial resources needed to improve capacity for contract negotiation have been met, in part, by the World Bank's Extractive Industries Technical Advisory Facility (EI-TAF). Nevertheless, the capacity for resource revenue administration, as well as for resource revenue management in general, remains low in many resource-rich developing countries, potentially leading to the loss of revenue.

Generating Extractive-Based Economic and Social Development

Following increases in commodity prices, the objectives of oil, gas, and mining sector policies have increasingly moved beyond mobilizing revenue. There is now a general consensus that resource policies need to provide for economic and social development beyond the end of oil, gas, or mineral extraction. Although there is no "one-size-fits-all" policy for resource-based economic development and industrialization, successful experiences reveal that strong government policies are necessary. Objectives of these policies may be, among others, to:

- Offset the depletion of finite subsoil assets with investments in human resources, infrastructure, and broader, long-term productive capacity and diversification.
- Leverage private sector investments in EI infrastructure to improve national public infrastructure capacity.
- Develop capacity for local, regional, and national value addition through the development of upstream and downstream linkages—as well as sidestream linkages to skills- and technology-based sectors and infrastructure—and support for entrepreneurs at the local and national levels.
- Develop human capital (in cooperation with resource companies) to address the staffing needs of companies, as well as those of related sectors—upstream, downstream, and sidestream.
- Ensure that resource extraction takes place in a manner that minimizes environmental degradation and promotes biodiversity.
- Ensure that benefits for communities, including owners and occupiers of land used for extractive activities, are defined through a fair process. It is important to note that the notion of fairness will change over the life of a mine due to (i) the difference between reality and the expectations held at the time of mine development, (ii) changes in commodity prices, and (iii) other changes (demographic shifts, external shocks—due to climate change, for example— and so on).
- Enforce health and safety standards.

Note

1. Most countries do not use contracts to regulate their relationships with foreign mining investors. Instead, they use mining code systems, in which most (if not all) of the rights and obligations of the investor are determined in the mining law or its regulations. While some highly developed economies use contracts and negotiations, these are most often used in countries that have relatively sparse mining legislation. The contract may be used as a transitional instrument that allows countries to participate in the mineral economy, while developing a more sophisticated and comprehensive system of mining legislation (and the state institutions necessary to support it). Most countries involved in contracts are located in Africa; contracts are also used in Central America, Central Asia, and Southeast Asia.

CHAPTER 2

The Economics of the Extractive Industries Sector

Accounting for Physical Stocks: Resources, Reserves, and the Economic Interpretation of Ore

"Resources" and "reserves" are common terms used throughout the extractive industries (EI), but in a way that can be confusing to those not familiar with the EI sector. Those not familiar with the terms might assume that "resources" are minerals or hydrocarbons that are available for depletion now, while "reserves" are saved for the future. In industry, the definitions take on very specific meanings, defined by codes (see appendix A for a discussion of the four main classification codes used for resources and reserves) that are then reinforced by legal requirements of disclosure for many publicly listed companies globally. Their definitions are somewhat contrary to intuition: mineral "resources" refer in general to a concentration of minerals of economic interest with the potential for eventual economic extraction, while a "reserve" is a portion of a resource that has proven itself to be legally, economically, and technically feasible for extraction. In that sense, reserves are more ready for depletion than resources, which have not yet satisfied the test of economic feasibility.

"Ore" is another term commonly utilized in the mineral industry, where, unlike "resources," it has both a technical and economic definition. In this context, ore is a portion of mineralized rock that contains sufficient minerals—with important elements, including metals—to make their extraction economically worthwhile. Not all ore bodies are created equal, and their economic value changes in response to external factors and policies. Ore bodies are not homogenous and contain varying qualities of ore which differ in an economic sense in relation to their proximity to the surface, the content of deleterious materials, rock hardness and ease of mining, and—very importantly—the concentration of main-, co-, and by-product metals or "grade." Because of this heterogeneity, the volume of an ore body can be sensitive to factors that affect economic returns, such as changes to prices, costs, fiscal policy, and technology.

The "cutoff grade" is critical to determining the boundary between ore and waste rock. If the concentration of metal in an ore body falls below the cutoff grade, extraction is not economically viable. More specifically, the cutoff grade is the minimum concentration of valuable product or metal that the mined material must contain before being sent to the processing plant. This definition is used to distinguish material that should not be mined or should be wasted, from that which should be processed (Rendu 2014).

The cutoff grade is not an exogenous feature of the ore body. Its definition may be affected by the operating and financing strategies of individual companies, economic or technical design constraints (such as equipment size or pit profiles and mining sequence), or technical performance criteria imposed by bank loans and other financial institutions.

The cutoff grade can increase or decrease, permanently or temporarily, and may have long-lasting impacts on the quantity of mineral resources that can be economically depleted from a mineral deposit. An increase in the cutoff grade termed "high grading" is a strategy that may sometimes be used by mining companies to increase short-term profitability (and the net present value of a project), thereby possibly enhancing returns to investors. However, increasing the cutoff grade is also likely to decrease the life of the mine by decreasing the level of economic reserves and also reducing the average grade of remaining resources and reserves. A change in the cutoff grade can occur due to factors outside management's control—such as a change in the country's fiscal policy. From a fiscal policy perspective, this is important: increased royalties, levies, and other fees increase the cutoff grade of a mine, similar to an increase in costs or a decrease in price.

Thus, a significant change to a royalty rate may lead to an increase in the cutoff grade and "high grading." In this instance, the firm's decision to increase the cutoff grade is an optimal response to policies set by a government that leads to the possible permanent sterilization of a portion of the nation's resources and a potential decrease in its subsoil wealth.[1] Furthermore, anticipation of such a policy may be enough to induce high grading, as the firm has a profit-maximizing incentive to deplete more resources while royalties are low, or otherwise risk depleting resources in the future, when royalties are higher and profits will be lower.

For example, the calculations in table 2.1 indicate the potential impact of a change (proposed in 2014) to the copper mining royalty rate for Zambia's open pit mines, to 20 percent from the existing level of 6 percent. The effect of the proposed royalty increase has the potential to increase the cutoff grade for Zambia's Lumwana-Chimiwungo resource to 0.31 percent copper (Cu) from 0.26 percent Cu. This would lower the overall copper resource available for mining and result in an estimated $700 million loss of in situ value (authors' calculations).

As illustrated by this example (see figure 2.1), high grading can result in a shorter mine life and less total resources depleted. This can reduce time-dependent opportunities such as those offered by price cycles and opportunities

Table 2.1 Preliminary Assessment of How Various Royalty Levels Would Affect the Cutoff Grade and Economic Feasibility of Zambia's Lumwana-Chimiwungo Resource

Assumptions (From the Barrick NI 43–101 Technical Report)

Direct operating costs:	$/t		
Average mining cost	3.76		
Average processing cost	9.72		
Average general and admin. cost	3.50		
	0% royalty	**6% royalty**	**20% royalty**
Total cost ($/t treated)	16.98	16.98	16.98
Cutoff grade (% Cu)	**0.25%**	**0.26%**	**0.31%**
Total cost ($/t Cu)	6,833	6,424	5,468
Total cost ($/lb Cu)	3.10	2.91	2.48
Price ($/lb Cu)	3.10	3.10	3.10
Royalty (%, gross)	**0%**	**6%**	**20%**
Net price ($/lb Cu)	**3.10**	**2.91**	**2.48**
Price–cost (breakeven)	0	0	0
Approximate tonnage above cutoff	447,360,574	436,324,806	400,773,623
Average grade above cutoff (% Cu)	0.65%	0.66%	0.69%
Copper contained in tonnage above cutoff (t)	2,894,423	2,865,491	2,761,428
Copper (lbs)	6,381,102,644	6,317,318,501	6,087,900,408
Copper production difference (lbs) (from 0% royalty)	0	−63,784,143	−293,202,236
Value difference ($) (from 0% royalty)	**0**	**−197,730,843**	**−908,926,932**
Copper difference (lbs) (from 6% royalty)		0	−229,418,093
Value difference ($) (from 6% royalty)		**0**	**−711,196,090**

Source: Authors' own calculations derived from Londono and Sanfurgo (2014). Courtesy of the Society for Mining, Metallurgy & Exploration (SME).
Note: Cu = copper; lbs = pounds; $/t = dollar per ton; $/lb = dollar per pound.

for spatial, forward, and backward linkages from the extractive industries. A shorter mine life can in turn reduce socioeconomic benefits such as long-term employment. Thus, there are many possible, unintended direct and indirect long-term consequences of changes to the cutoff grade.

As the strategy of high grading illustrates, the notion that reserves are a fixed stock is mistaken; private sector strategies, public policies, and external factors all have an impact on a nation's subsoil wealth in the short and long run. Operating strategies and public policies can affect the total stock of resource available for depletion, and this will in turn affect costs through optimal-scale economies—thus, ore and reserves are endogenous to fiscal policy and are not fixed and exogenous, as is often assumed. (This endogeneity is particularly evident when the cutoff grade is changed.) In sum, ore and reserves are not fixed in volume or value but vary continuously over time in response to changes in policy, prices, and costs.

Figure 2.1   Graphical Representation of How a Change in Royalty Would Affect the Cutoff Grade and Economic Feasibility of Zambia's Lumwana-Chimiwungo Resource

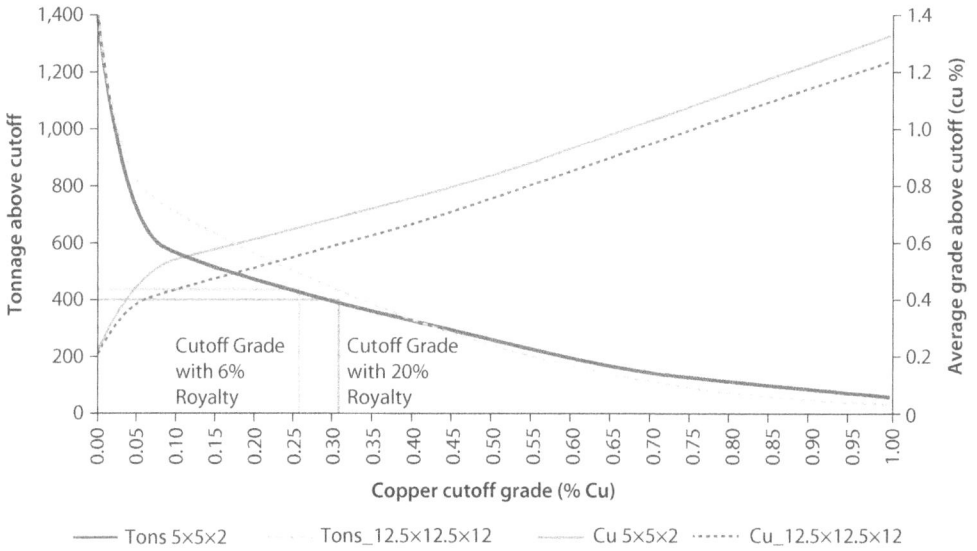

Source: Barrick Gold Corporation estimates, derived from Londono and Sanfurgo (2014).
Note: The effect of the proposed royalty increase has the potential to increase the cutoff grade at the Lumwana-Chimiwungo resource to 0.31 percent Cu from 0.26 percent Cu. Based on the grade tonnage relationship for the deposit, the increased cutoff grade decreases economically feasible resources by approximately 35 million tons or by approximately 230 million lbs of copper. At a price of $3.10/lb, the lost in situ value to the Chimiwungo resource alone due to the royalty is approximately $700 million. Cu = copper.

Theory of Rents and Valuation of Subsoil Assets

As can be seen in figure 2.2, despite the fact that Zambia is known for its high-grade deposits, copper mines in its Copperbelt province are less competitively positioned and more vulnerable to changes in prices—when considering the high costs of extraction and transportation—than in some other regions of the world. Where mines for bulk commodities (for example, iron ore, coal) are located near the earth's surface and in easy reach of maritime transport, production costs can be significantly lower. Some mines are also linked to smelters that have low energy costs, or have high availability of skilled labor. Other deposits are not quite so easy to extract but are still profitable. Prices correspond with the marginal cost of the marginal producer; because most minerals are perfect commodities (that is, one pound of copper can be easily replaced by another),[2] sales prices are similar for all producers. This results in low-cost producers earning potentially significant economic rents.

Economic rents are any payments to a factor of production in excess of the cost (including capital and returns to capital) needed to bring that factor into production. Economic rents are not confined to the oil, gas, and mining sectors but exist wherever there is a fixed factor of production. As in those sectors, several different types of rents are present in the extractive industries: Ricardian, Hotelling, and quasi-rents (such as those associated with returns to capital).

Figure 2.2 Cost Curve of Copper Mine Production, Selected Projects, Zambia

U.S. cents per pound

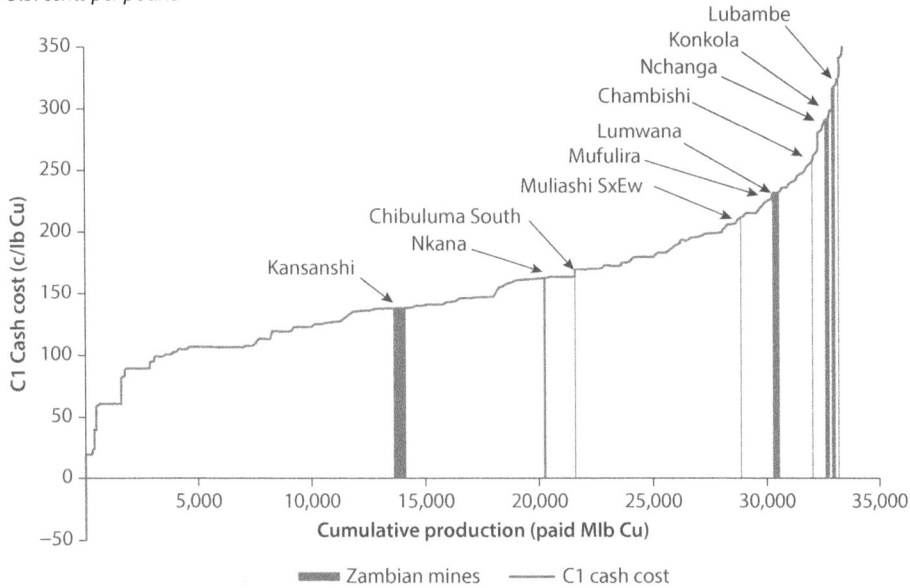

Source: Wood Mackenzie 2013.
Note: c/lb = U.S. cents per pound; Cu = copper.
1. Each of the columns highlighted in blue represents a single producer of different quality. The height of each column reflects the costs of producing a given unit of copper or another metal. Those in blue represent Zambian producers. Negative costs represent the production of by-product copper. This may represent cases where copper is associated with another metal (such as gold or cobalt) and the production of this metal subsidizes the copper production, resulting in negative copper production costs.
2. The C1 cash-cost curves are a measure of all direct costs, expressed in U.S. cents per pound ("c/lb") on a "paid copper"; C1 costs can be interpreted as short-run costs. Net direct cash costs (C1) include the cash costs incurred at each processing stage up to delivery to the market, less net by-product credits if any. These costs include mining, ore freight, and milling costs; mine site administration and general expenses; concentrate freight, smelting, and smelter general and administrative costs; and marketing costs.
3. The positions on this chart represent a "snapshot" of industry supply. The position of individual mines is expected to vary over their life cycle.

Ricardian rents cannot be diminished through competition. Since they are generated by the resource itself and not entirely by the application of skill or experience, they can be taxed without distorting optimal decision making. Hotelling rents are the opportunity costs of producing one more unit today instead of in the future; quasi-rents reflect the return to capital and other fixed costs. Because both Hotelling and quasi-rents are real costs, taxing them would result in distortionary production decisions. (See figure 2.3 for an illustration and appendix B for more on all three types of rent.)

Valuation of Subsoil Assets

The value of a nation's subsoil assets is the sum of the economic value of reserves and other resources. The value of reserves should include three values: Hotelling rents,[3] Ricardian rents, and option value. The value of other resources should include the option value only. These values will be reflected in market values for reserves and resources when markets are functioning well—that is, when there are

Figure 2.3   Conceptual Depiction of Ricardian and Hotelling Rents

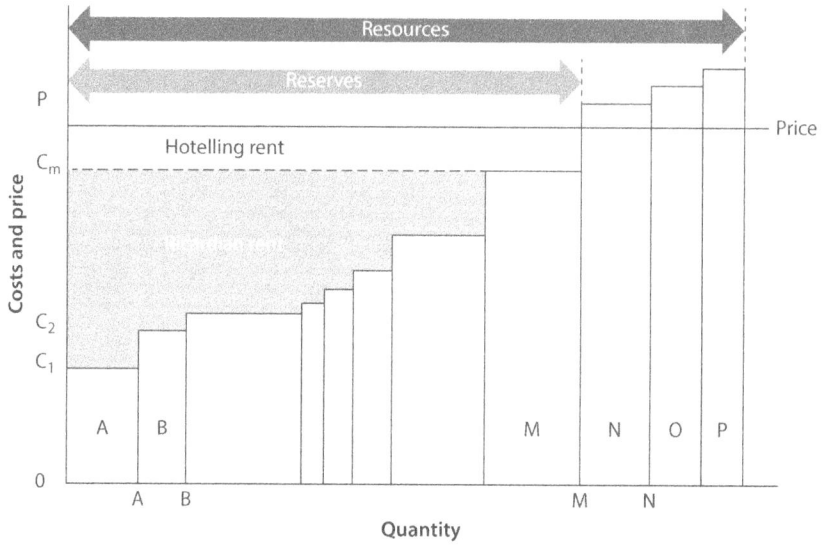

Source: Derived from Nordhaus and Kokkelenberg (1999) and Otto and others (2006). Reprinted with permission from the National Academies Press, copyright 1999, National Academies of Sciences.
Note: Each of the columns A through P represents a single actual or potential producer of different quality. The height of each column reflects the costs of producing a given unit of copper or another metal. For example, for the most productive mine or for the mine with the highest grade, the mine in column A, the costs of production are $0C_1$. For the mine in column B, the costs of production are $0C_2$. Cm represents the cost of the high-cost mine in column M, which generates no Ricardian rent but where the cost-price differential is still sufficient to cover the opportunity cost or Hotelling rent, hence the mine will produce. The amount of metal that each mine can produce is given by the width of its column. So the most productive mine, column A, can produce the quantity 0Qa of metal, and the mine in column B the quantity QaQb. P is the market price when demand requires metal production from mines in columns A through M when user costs, or Hotelling rents, exist. Mines N, O, and P are not economic given those prices. Those producers with a marginal cost of production below the market price would be expected to classify resources that could be produced at their costs within "reserves," while those with a marginal cost above the market price would classify those resources as "resources."

many potential buyers and sellers of reserves and resources, and there is good foresight about future technological and market conditions, good knowledge of subsurface assets, and generally rational decision making (Nordhaus and Kokkelenberg 1999). In some cases, rents associated with imperfect market structures exist and should be included in the evaluation of reserves and other resources. For example, this is necessary where a long and stable market power exists, as is the case for oil (the Organization of the Petroleum Exporting Countries, OPEC), potash (regional marketing networks), and, until recently, diamonds (De Beers).

Techniques for valuing subsoil assets are of three types, based on the market (such as transaction value and ratio analysis), cost, or cash flow (such as net present value, NPV).[4] These approaches are described briefly below:

- *Market based.* The transaction value is commonly used to evaluate human-made physical assets in national accounts and sometimes to support valuation ranges in cases of corporate mergers and acquisitions. Using this approach, the value of subsoil assets is based on actual transaction values, as when one company sells an asset (or group of assets, such as a corporation) to another company. To derive the underlying value of the subsoil asset, however, requires adjusting a transaction price for other values embodied in the price, such as cash on account, associated capital (roads, processing facilities, and so on), royalty and tax obligations to governments, and other company-specific assets or liabilities. It must be noted that transactions may be infrequent or not reported in the public domain, making them difficult to use as a basis for valuation.

- *Cost based (for example, using replacement costs).* The understanding behind this approach is that markets provide incentives to invest in replacing a depleting subsoil asset up to the point that the marginal or additional cost of finding the replacement equals the market value. This method, however, is only valid for valuing resources of the same quality and therefore ignores a major source of rents—the differential quality of subsoil assets.

- *Cash-flow based, using NPV.* This is the most widely used standard method for valuing subsoil assets. It requires defining a stream of expected future revenues and costs, and then discounting them, using an appropriate discount rate. As such, using NPV requires information on subsoil asset quantities, qualities, current and future costs of production, current and future prices, and the discount rate. A special case of the cash-flow-based method is called Hotelling valuation, sometimes referred to as the net-price method. In this case, the NPV of a subsoil asset is simply the net price at present (revenues less all costs), multiplied by the size of the resource stock. There is no need to discount explicitly, because if the asset is being extracted efficiently, and certain other conditions hold, then its value will grow over time at the rate of interest, exactly offsetting the discounting of future values at the same rate. This method is convenient because it requires neither determining the appropriate discount rate nor forecasting future prices and costs; resulting assumptions, however, risk being unrealistic (Nordhaus and Kokkelenberg 1999; Statistics Canada 2006).

Guidelines for the valuation of subsoil assets suggest that the appropriate technique for valuation varies with the level of geological information. Examples of two guidelines used for the valuation of mineral properties include the Canadian Institution of Mining Valuation Guidelines (CIMVal)[5] and the South African Code for the Reporting of Mineral Asset Valuation (SAMVAL).[6] A common feature of these codes is that the acceptable valuation technique differs depending on the nature of the property or the use of the valuation. For example, the cost-based approach may be acceptable for valuing an exploration property where little information is known, but is unacceptable for valuing operating mines.

Structure of Energy and Mineral Markets

A comprehensive analysis of mineral and energy markets requires application of basic economic principles, complemented by an understanding of institutional arrangements (from exploration to beneficiation) and the technology of extraction. The following draws heavily from the works of John Tilton, a prominent mineral economist, who writes:

> Studies by good economists who apply their theoretical concepts in ignorance of important technological and institutional constraints are almost inevitably sterile and misleading. The same can also be said for commodity specialists, who may know well the relevant institutions and technologies but who lack a basic understanding of economic principles. *Good analysis requires knowledge of both economics and the particular metal of interest.* (Tilton 1985)

This section provides a brief overview of the different building blocks of mineral and energy markets; the reader is referred to Tilton (1985) for a more comprehensive discussion of the subject.

Time Frame for Analysis

Distinguishing between four distinct time periods is useful for framing economic problems. When designing their models, economists first define the time frame of the analysis. Considerations that may be critical for a short-term analysis may not be so for a long-term one. Any analysis of energy and mineral markets should be defined by one of four distinct time periods:

- *The immediate*, when all production is fixed and supply adjustments are principally achieved through changes to inventories.
- *The short term*, when, on the demand side, market conditions are considered fixed, and on the supply side, production capacity is fixed. Supply-side adjustments are realized through the utilization of existing capacity.
- *The long term*, when no constraints on the demand side are assumed, but the supply side is limited to the depletion of known deposits and to existing technologies. Adjustments in the long term are constrained by investment in new productive capacity.
- In the *very long term*, both demand and supply are entirely unconstrained.[7] Metal and energy supply adjustments in this time frame can be achieved through exploration and research and development.

This framework is summarized in table 2.2.

Table 2.2 Summary of Constraints to Demand and Supply across Time Periods

Time period	Demand	Supply	Adjustment (supply)
1. Immediate run		Production is fixed	Inventories
2. Short run	Market conditions are considered fixed	Capacity is fixed	Capacity utilization
3. Long run	Unconstrained	Limited to known deposits and existing technologies	Constrained investment
4. Very long run		Unconstrained	Exploration and R&D

Source: World Bank, courtesy of the Society for Mining, Metallurgy & Exploration (SME).
Note: R&D = research and development.

In the next subsections, the nature and structure of demand and supply will be discussed in more detail.

Determinants of Mineral and Energy Demand

The demand for mineral and energy products is principally determined by (i) the ability and willingness of buyers to pay and (ii) derived demand, since many mineral and energy products are intermediate goods in manufacturing processes. It should be noted that the determinants of demand vary by commodity; in the interest of space, this study focuses on common determinants of demand across all mineral and energy markets.

There are seven determinants that are critical to understanding the structure of demand across all energy and mineral markets:

- *Income and economic activity.* Mineral and energy products are most often consumed as intermediate products used in the production of consumer goods; changes in the demand for these final goods have a direct impact on the demand for raw materials.

  In turn, the demand for final goods is a function of income or economic activity, which can be affected by two factors. The first factor is normally short-term changes that occur largely as a result of fluctuations in the business cycle. The second factor is longer-term changes driven by secular growth and structural change in the economy (often conceptualized as the "intensity of use," IoU). Whichever of these is primarily responsible for an income change will affect the magnitude of the demand response. (The longer-term structural dynamic of energy and mineral demand, including the IoU, is discussed in detail in appendix C.)

- *The price of the mineral or energy product.* In addition to income, a mineral or energy commodity's own price is a key determinant of demand. A higher price increases the cost of production of the final good, and if this cost is passed through to the price of the final good, all else being equal, the quantity demanded of the final good will diminish. If a cost-effective substitute can be found, manufacturers using the raw material as an input are motivated to shift away from the higher-priced metal.

- *The price of substitutes.* Most metals used as intermediate products compete with other materials (including other metals) that provide a similar function. For example, in the fabrication of pipes for cooling or cooking, aluminum competes with copper.

  The relationship between the aluminum price and copper price is well established, and analysts trading in those commodities regularly check the ratio of one price to the other (see figure 2.4). This is because aluminum and copper provide similar thermodynamic properties and are thus very good substitutes for use in radiator piping and cooking applications. There is therefore a copper–aluminum

Figure 2.4 Three-Month Copper Prices Compared with Three-Month Aluminum Prices, 1990–2012

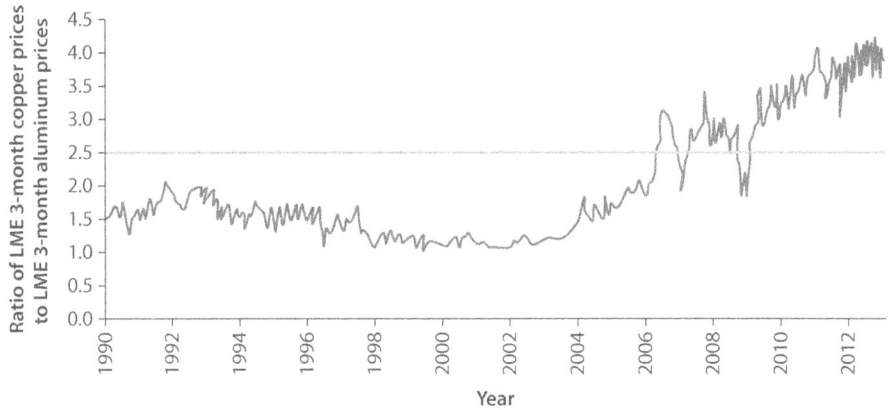

Source: *2014 Metal Outlook: Exploring Bull and Bear Risk Factors*, J.P. Morgan Commodities Research (Kaneva 2014).
Note: LME = London Metal Exchange.

price multiple; above this, buyers are incentivized to switch to the other one of the two. Market players monitor this multiple closely to understand the potential risks of particular metal contracts.

Substitutes in energy markets are obvious in some instances; electrons transmitted down a power line are perfectly substitutable. In certain circumstances, renewable energies such as solar or wind power can be a near-perfect substitute for diesel or coal power.

- *The price of complements.* In some instances, the fall in the price of one material will increase demand for another. For example, a decrease in the price of iron ore (a major input in steel manufacturing) may increase the quantity demanded of steel, but may also increase the demand for all other metals used in alloying different types of steel. In this case, iron ore is a complement to alloying metals.

- *Technology.* Technological change affects demand in several ways:
  - Through increased efficiency of use—such as when innovations in the manufacturing or recycling of aluminum cans decrease demand for bauxite, the ore from which aluminum is extracted
  - Through eliminating the entire demand for a commodity—for example, the innovation of polyvinyl chloride (PVC) piping has significantly reduced demand for copper in plumbing
  - By creating entire new markets—such as those created by investment in photovoltaic (PV, or solar) panels that require the use of scarce metals such as indium, selenium, tellurium, and gallium

- *Consumer preferences.* Changes in consumer preferences ultimately alter the end-use markets for commodities and thus have a variety of possible effects, including increasing or decreasing demand (both in the long and short terms) and affecting elasticity. Consumer preferences are market specific and, in some cases, cultural.

For example, in the case of diamonds most of the world's demand resides in the United States and Japan, where consumer preferences are very different. In a market where a diamond is graded against the four Cs (color, carat [weight], cut, and clarity), the U.S. consumer will tend to prefer carat (weight) over the other Cs because diamonds are associated with wealth, and bigger is better. Japanese consumers, on the other hand, tend to prefer those features that might be interpreted as representing purity—such as color and clarity—over size.

- *Government policies.* Government policies, regulations, and other actions also affect demand. Key examples include foreign, industrial, and trade policies. In the case of industrial or development policy, a government may invest in infrastructure that creates demand for steel, copper, aggregate materials, and various complements. In an example of trade policy, stockpiling strategic reserves can create short-term, demand-side effects as countries either build up or sell down the reserves. For example, gold reserves were destocked on a large scale when central banks moved away from the gold standard; today, both the U.S. military and the Japanese government keep stocks of certain minerals they define as critical.[8]

Elasticity of Demand

The relationship between demand and its major driving factors can be expressed by the generalized (short-term) demand function in equation 2.1, where demand depends on income (Y), own price (P^O), the price of substitutes and complements (P^S and P^C, respectively), technology (Tt), consumer preferences ($Cons$), and government policies (Gov) during period t.

$$Q_t^D = f\left(Y_t, P_t^o, P_t^s, P_t^c, T_t, \mathrm{Cons}_t, Gov_t\right) \qquad \text{(equation 2.1)}$$

Income elasticity of demand is elastic in the short term. Also in the short term, the income elasticity of demand for most mineral products is assumed to be greater than 1, since demand is concentrated in the construction, transportation, capital equipment, and the consumer durable sectors, where demand tends to be highly responsive to the business cycle.

Demand is more price elastic in the long run than in the short run. The elasticity of demand with respect to a metal's own price (as well as the prices of substitutes and complements) is normally assumed to be greater in the long run than the short run. This is because it often takes time—several years or more—for metal and mineral consumers to change production technologies and substitute one material for another (Tilton 1985). A significant part of the impact of a price change on metal demand occurs over the very long run as an indirect result of new innovations and technologies induced by changes in

Figure 2.5 Illustrative Demand Curves in the Immediate, Short, Long, and Very Long Run

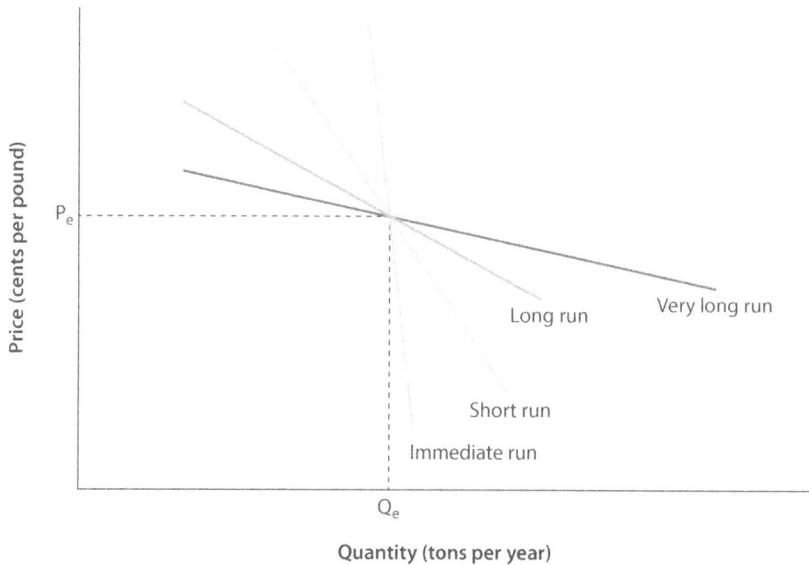

Source: Tilton (1985), courtesy of the Society for Mining, Metallurgy & Exploration (SME).

material prices. Thus, own price elasticity of demand is normally presumed to be less than 1 in the short run and greater than 1 in the long run. The short-run demand curve will normally have a steep slope (Tilton 1985), much steeper than the demand curves for the long and very long run, as shown in figure 2.5.

When considering the elasticity of demand for minerals that are used in manufactured or compound goods, the Hicks-Marshall laws provide a useful framework. These state that, in general, the elasticity of demand for a mineral or energy product will be high (that is, formally $|\varepsilon_D|>1$) when (i) the own price elasticity of demand for the final produce is relatively high (that is, formally $|\varepsilon_{\text{fin prod}}|>1$); (ii) when it is easily substituted; (iii) when the own price elasticity of supply of other production inputs is high (that is, formally $|\varepsilon_{\text{S prod factors}}|>1$); and (iv) when the mineral or energy products account for a large percentage of the total final costs of the final product.

Determinants of Mineral and Energy Supply
Mineral and energy supply can be subdivided into primary and secondary supply. The nature of mineral and energy supply varies greatly between *primary production* (direct from newly depleted deposits or reservoirs) and *secondary production* (the recycling of consumer goods or manufacturing waste).

Primary supply can be further categorized as main, co-, and by-product supplies. The category depends on the importance of the mineral or energy product to the economic feasibility of the extractive project. For example, main products are so critical to the viability of a project that their price alone determines the project's output. When the prices of two or more joint products affect the level of production, these are called coproducts. Finally, a by-product is almost entirely unimportant: its price or recovery has no impact on the amount of production from the mine.

Secondary supply can be a major source of supply and—together with main, co-, and by-products—needs to be considered in any supply–demand assessment. Iron ore, bauxite, and coal are typically produced as individual products and are thus most often considered to be the main product supply for iron, aluminum, and coal, respectively. But since iron, steel, and aluminum produced from iron ore and bauxite are often recycled from consumer products, secondary production can compose a large portion of total supply for these metals. Copper, on the other hand, though often found with other metals (such as gold or cobalt), is often a main product. As in the case of iron and aluminum, secondary production can sometimes be a significant component of total supply. Lead and zinc, meanwhile, are often coproducts—they naturally occur together in deposits. While secondary production is not important for zinc, the recycling of lead in automobile batteries is a very significant source of lead. Gold, silver, molybdenum, and tellurium are typical by-products of copper production, while gold and silver are by-products of nickel, lead, and zinc production (and are main products in some mines).

The distinction between main, co-, and by-product production is important, because it relates directly to the marginal cost of extracting these products. Where main product supply is critical to a mining, oil, or gas operation, it bears the costs of all the direct variable expenses and fixed expenses associated with depletion. Where a metal exists as a coproduct, costs are shared with the other coproducts; where it is produced as a by-product, most of the costs of extracting are borne by the main product. Thus, by-product supply tends to be the lowest-cost supply. For example, when considering a metal, such as gold, that may be produced as any one of the three product types, it is likely that main product supply will cost more than coproduct supply and that by-product supply (say, gold production associated with copper production) will be the lowest-cost option.

It is also important to distinguish between short- and long-term supplies, as supply is highly inelastic in the short term. This is because in the short term most EI operations face a capacity constraint; regardless of what happens to prices, capacity is fixed at its installed level. Companies can change production through more efficient use of capacity; however, given the large fixed costs associated with the EI sector, it is optimal for extractive operations to run continuously year round, leaving very little room to increase capacity. Thus, in the short run, supply is highly inelastic.

Price volatility in the metals market is primarily a result of the high income elasticity of demand, combined with the low price elasticity of supply. Because of the low elasticity of supply, a small change in demand can have a significant increase in price (as supply is not able to adjust). Given the long lead times to bring new production on stream, price increases can be significant—resulting in the "boom-bust" cycles often associated with commodities markets.

In the long run, new mines can be developed and new processing facilities can be developed or expanded with the application of capital. As a result, supply is considerably more elastic in the long run than in the short run or immediate future.

In the very long run, no constraints exist with respect to known deposits or production facilities: firms have time to apply capital and look for new deposits. The exhaustion of existing types of new deposits can allow for the depletion of new, higher-cost resources that are either deeper or require higher risk-adjusted returns to capital. As a result, in the very long run the supply curve becomes horizontal, and there are effectively no constraints on supply.

High price levels can sometimes increase the elasticity of supply as different types of deposits become feasible. For example, large, low-grade copper deposits are fairly abundant and have been found frequently over the past 70 years. A study recently conducted by the United States Geological Survey, USGS (Zientek and others 2014), has determined that Zambia contains approximately as much undiscovered copper as discovered, but the new copper is either low grade (for example, porphyry deposits) or deep seated. As the price of copper rises and technology improves, these types of deposits become feasible. In another example, gold can be found dissolved in the world's oceans in large quantities, but at low concentrations. At a price threshold, this gold will become economically feasible to extract.

The relative positioning of the supply curves in figure 2.6 is based on several assumptions regarding the influence of production costs on supply. For example,

Figure 2.6 Illustrative Supply Curves in the Immediate, Short, Long, and Very Long Run

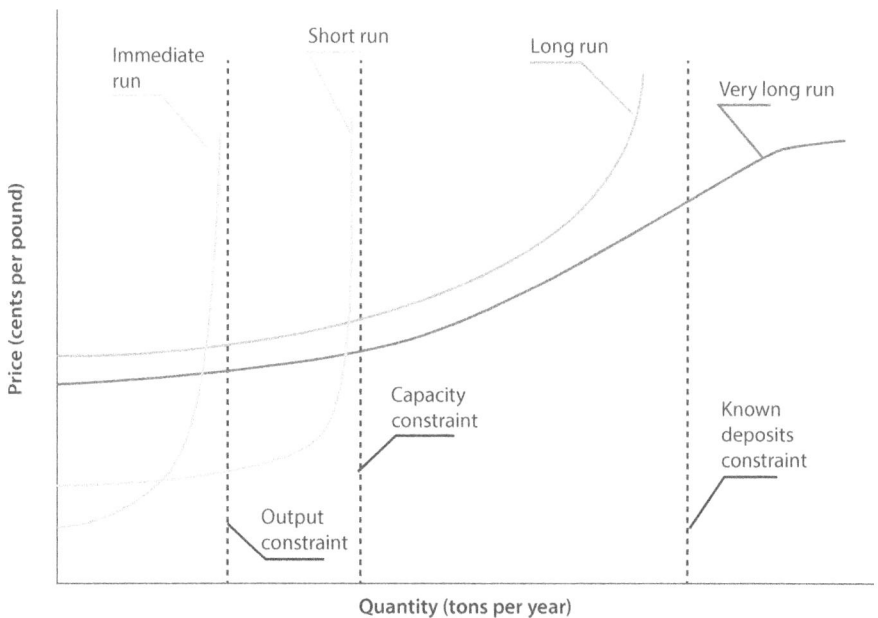

Source: Tilton (1985), courtesy of the Society for Mining, Metallurgy & Exploration (SME).

because of the fixed nature of production in both the immediate and short run, these costs are likely to be similar (Tilton 1985). Immediate supply relates to inventory and not to new production, and as a result, costs are slightly lower. In the long run, new and higher-cost deposits are presumably going to be brought into production, and so costs are assumed to increase (as existing lower-cost deposits will already have been extracted). The short-run curve lies below the long-run curves, because it only includes short-run costs and does not include fixed costs associated with capital. The long-run curve is shown above the very-long-run curve because in the very long run the discovery of low-cost deposits and the development of new deposit types should reduce costs—and, in turn, the price required to incentivize any level of supply.

As with demand, many variables affect supply in the short and long run, in a way that varies by commodity. The following discussion examines six key factors (see Tilton [1985] for details):

- *Own price.* Profit-maximizing firms will increase production until marginal costs equal marginal revenues. As previously discussed, the capacity constraint in the short run limits the expansion of supply; however, in the long run, firms are highly responsive to price signals. Not only do production patterns closely resemble (but lag) most commodity prices, exploration activity (a proxy for long-run supply) is also responsive to prices (as shown in figure 2.7).

- *Input costs.* Because key input costs (such as labor and energy) directly affect the total cost of production, they can have a material impact on supply. In the

Figure 2.7 World Gold Exploration Expenditures versus Gold Prices, 1975–2012

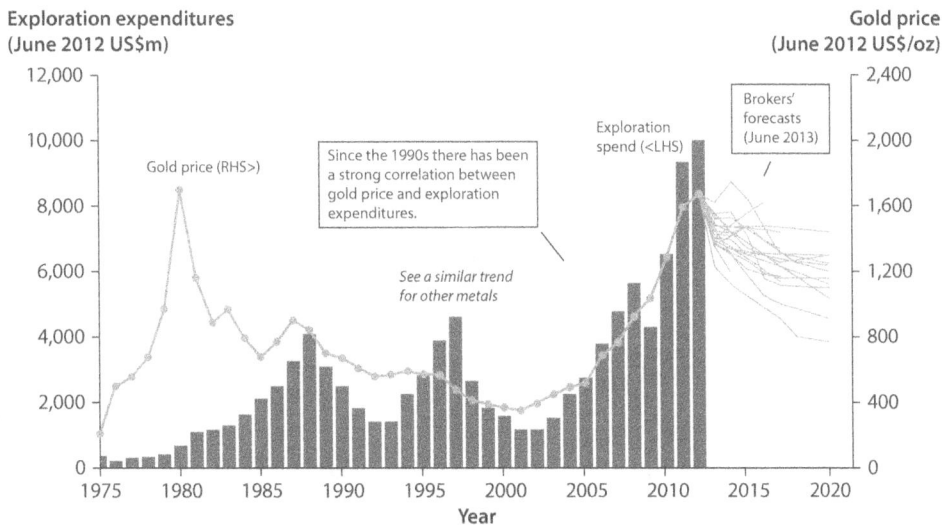

Source: MinEx Consulting Pty Ltd. estimates, 2013.
Note: RHS = right-hand side, "Gold Price"; LHS = left-hand side, "Exploration Expenditures."

short run, because of a low elasticity of substitution among factors of production, a price increase in one factor of production (such as labor) will often
directly translate into a direct shift in a producer's own supply curve. In the
long run, firms can adjust technologies and design the optimal mix of various
factors, given these factors' current prices and projected future prices.

- *Technology.* Advances in technology can have a wide-ranging effect on mineral
  and energy supply. In some cases, the effect can be dramatic, unlocking significant resources that were previously subeconomic. Consider the development of solvent extraction and electrowinning technology for the processing
  of low-grade porphyry copper, first in the United States and later in Chile
  (where its use is now widespread).

- *Labor and other disruptions.* Labor disputes and other disruptions—political,
  technical, and so on—can have a significant impact on mineral supply. Pro-
  tracted labor disputes in platinum production mines in South Africa (from
  2012 to date) provide one example. To avoid disruptions in the long term,
  some producers have considered substituting labor with capital in the form of
  specialized mechanical equipment. Copper mine supply disruptions in 2014
  exceeded 2013 and 2012 levels. Reduced operating rates in Indonesia and the
  slower-than-expected expansion of several other operations contributed to
  unplanned losses and pit instability at Bingham Canyon in the United States.
  Total unplanned disruptions in mining operations were associated with output reductions of 673,000 tons in 2014 (equivalent to ~75 percent of all of
  Zambia's copper production) and were due to a range of issues, from export
  bans (Grasberg, Indonesia; 150,000 tons) to delays in securing power
  (Toromocho, Peru; 90,000 tons) to technical problems such as rock falls and
  subsidence (Mount Lyell, Tasmania; 24,000 tons; and Birla Nifty, Australia;
  26,000 tons) (table 2.3).

- *Government activities.* Policy and regulations can have a significant impact on
  production costs and the development of new supply. Examples of influential
  factors include frequent changes to key policies such as those governing tax,
  uncertainty regarding announced policy intentions that have not yet been
  formulated into laws and regulations, mineral export bans, and the national-
  ization or indigenization of oil and mining companies.

  These actions affect supply by (i) putting direct, physical limits on the amount
  of supply available for global markets (that is, through export restrictions such as
  quotas or duties); (ii) limiting incentives for new supply (though price ceilings);
  (iii) discouraging investment in new productive capacity (though frequent policy
  changes or unclear policy announcements); and (iv) possibly constraining sector
  investment and productivity (through direct state control of operations).

  Policy incentives such as the provision of state guarantees and tax holidays,
  meanwhile, aim to stimulate new supply. In most developed countries, these

Table 2.3 Selected Copper Supply Disruptions in 2014

thousands of tons

Operation	Mine loss	Smelter loss	Notes
Mt Lyell	24		Rock fall
Toromocho	90		Delays securing power supply, water discharge concerns, "equipment-related issues"
Caserones	75		Construction snags, strike
Grasberg	150		Ore export ban, accident
BatuHijau	18		Ore export ban
KCM	18	16	Mine: stoppages, low grades, flooding. Smelter: blending challenges, "commercial issues"
DRC	50		Power rationing, power stability
OyuTolgoi	25		Technical problems, rake failure in tailings thickner
Nifty	26		Subsidence
Ventanas		24	23-day closure for environmental-related repairs
Los Bronces	3		Contractor strike
Antapaccay	1		Repairs after typhoon damage inflicted last year
Pasar		55	Mainly partial conveyor collapse at Lumwana
Barrick various	27		Start-up teething problems
Ministro Hales	10		Further start-up delay
JabalSayid	55		Intersection of geological fault
El Soldado	14		Low ore grades reported
MantosBlancos, Mantoverde	3	3	Fire in cooling tower
LS-Nikko No. 1		2	Explosion in blast furnace
LS-Nikko No. 2		18	Faulty oxygen generator
Gansu		120	Low grades
Antofagasta various	14		Ball mill failure
Pinto Valley			Strike
RadomiroTomic	48	48	Toxic spill
Mt Polley	7.5		Scrap shortages
GuelbMoghrein	5		Strike
Buenavista			Sequencing, accident
Various		40	
Escondida	7		
Chelopech, Kapan	2		
Total Losses	673	326	

Source: Metal Bulletin Research (MBR) (Westgate 2014).
Note: DRC = Democratic Republic of the Congo; KCM = Konkola Copper Mines.

incentives are limited to certain critical or strategic minerals. These most often relate to energy but may also include rare minerals critical for security reasons or domestic industries. Fiscal incentives need careful evaluation by governments before being offered. (The authors are not advocating the use of incentives, which have yielded mixed results for both countries and firms. Instead, the authors merely seek to point out that incentives may form part of the supply function by shifting the firm's cost of supply.)

- *Market structure.* Finally, market structure can have a significant impact on supply, particularly where cartels or oligopolies exist (such as in the oil and potash markets).

Elasticity of Supply

The relationship between supply and its major driving factors can be expressed by the generalized function in equation 2.2. Here, the quantity demanded depends on own price *(P^O)*, costs (C$_t$)—which can be further broken down into wages, energy, and capital services—and technology *(T$_t$)*, the natural resource characteristics at any given time *(R$_t$)*, and government policies *(Gov)* during period *t*.

$$Q_t^s = f\left(P_t^o, C_t, T_t, R_t, Gov_t\right) \qquad \text{(equation 2.2)}$$

The supply function is simplified: it does not include labor or other disruptions, or any lagged values for prices or costs; also, it is a short-run supply function.

The own price elasticity of supply is generally greater in the long run than in the short run, but this is true only when output in the short run is approaching capacity (Tilton 1985). When capacity is underutilized, supply can respond to changes in price even in the short run.

Production Technology and Long-Run Supply

Extractive industries seldom correspond to a Cobb-Douglas production function. It is worthwhile to touch on the important issue of production technology. Economists tend to apply Cobb-Douglas or constant elasticities of substitution (CES), but these seldom reflect industrial reality in the short run. If a mine is functioning at capacity, an increase in laborers does not necessarily increase throughput. Mining, oil, and gas ventures require inputs in fixed proportions in the short run. One truck needs one driver, not two. A second driver does not make the truck twice as productive. Similarly, more trucks do not increase the capacity of a treatment plant. Thus, in the short run, most inputs in EI are fixed in proportion to one another.

Short-run production technology in an EI setting is more accurately described by Leontief production technology than the Cobb-Douglas or CES technologies. This is an important realization: it recognizes the significant limits on the extent to which factors of production can be substituted for others and, perhaps more important, that the marginal cost will also be the average cost. This is important because it gives weight to industry cash cost curves of supply (such as in figures 2.1 and 2.2), where what is most likely being reflected by the C1, C2, or C3 terminology is not marginal C# costs ($\forall$ #$\in$ *{1,2,3}*), but rather the average C# costs over a given period.

As with the price elasticity of supply and demand, the elasticity of the substitution of factors of production is significantly higher in the long run. Thus, in the long run, CES, Cobb-Douglas, or other nested functions form a suitable basis for models (see Miller [2000] for more on how to integrate long-run thinking with short-run realities).

Notes

1. It should be noted that not in all cases will high grading lead to sterilization. In some cases, high grading may simply lead to a deferment of producing the low resource. This is more likely in open pit mines or in reworking old tailings, as some of the safety and access conditions, albeit potentially still significant, tend to be easier than in underground workings.

2. Precious stones are an exception to the perfect commodity model. One carat of diamonds, for example, cannot be replaced by another; there are approximately 12,000 categories used to assess rough diamond production.

3. Whether Hotelling rents should be included in the valuation is a subject for discussion, since Hotelling rents represent a real opportunity cost of current production. Practical considerations, however, force their inclusion in most valuations of subsoil wealth, because separating them from Ricardian and quasi-rents can be difficult.

4. See Nordhaus and Kokkelenberg (1999), on which the discussion here is based.

5. http://web.cim.org/standards/documents/Block487_Doc69.pdf.

6. http://www.samcode.co.za/.

7. The depletion of fixed stock natural resources may be defined as geological or economic. This analysis assumes economic depletion.

8. U.S. military: Ga, Li, Nb, rare earths, Re, Ta (Parthemore 2011); Japan: Co, Cr, Mn, Mo, Ni, W, V (Eggert 2011).

The Extractive Industries Sector • http://dx.doi.org/10.1596/978-1-4648-0492-2

CHAPTER 3

Institutional Framework

Mandates and Coordination

The institutional framework for extractive activities should be based upon clear constitutional and other legal mandates for the legislature and the executive. The framework should clearly outline the distribution of responsibilities across government and provide checks and balances on the level of discretion to be had by individual government entities:

- *Legislative bodies*, as part of their law-making mandate, are responsible for reviewing bills and enacting legislation for the extractive industries (EI) sector. Through their oversight function, legislatures have the mandate to ensure the accountability of government activities and the allocation of funds.

- *The executive*, typically including the presidency and the executive cabinet, often makes the final decisions on critical issues such as licensing, state participation, and the establishment of EI stabilization and/or savings funds.

The formulation and implementation of policies for the EI and related sectors is a highly complex task (see table 5.2 on the components of an EI sector program) that involves a number of ministries and government agencies. These units in turn need to interact with companies, civil society, and other actors in the sector. This requires coordination across ministries, as well as specialized human resource capacity within the government.

Interministerial coordination is critical to prevent overlapping or conflicting roles and to avoid gaps in regulatory responsibility. Among many possible examples of insufficient coordination, it is not unusual to find that one ministry or agency promotes investment in the EI sector even as others degrade the business environment by creating barriers such as delays in issuing visas and work permits, and in customs clearance for goods and equipment. Box 3.1 highlights the case

---

Box 3.1 Insufficient Institutional Coordination and Its Impact: The Case of Ghana

A recent report by the Ghanaian Public Interest and Accountability Committee (PIAC), a statu-
tory body charged with monitoring and evaluating compliance with the Petroleum Act by
the government, highlights some challenges in petroleum contract management, arising
from what the report considers to be insufficient institutional cooperation:

- *Overoptimistic revenue projections.* The Ministry of Finance and Economic Planning
  (MOFEP) that was in charge of the forecasting methodology for petroleum revenue did
  not, according to the report, sufficiently consider the advice of the Ghanaian Revenue
  Authority (GRA) with regard to the tax-paying position of the companies involved. As a
  result, actual oil revenue was almost 50 percent less than the budget forecast.
- *Fragmented accounting of revenue streams.* All revenue receipts from petroleum are legally
  directed through the Ghana Petroleum Holding Fund, which allows for consolidated
  reporting and monitoring. But surface rental fees are paid into a separate nontax revenue
  account, which may have resulted in some undercollection.
- *Insufficient capacity and lack of budgetary provision for agencies.* Some of the agencies
  required to enforce and monitor petroleum contracts and manage petroleum revenues,
  as mandated by law, are not sufficiently staffed and are not provided with adequate
  resources through the budget. Insufficient staff and resources for the technical audit of
  operations are identified as a particular weakness—and a key factor behind the improper
  enforcement of operational commitments for companies engaged in crude oil exploration and production in Ghana.

Source: Adapted from the Ghana Public Interest and Accountability Committee (2012).

---

of Ghana, where insufficient coordination is reported to have impacted resource
revenue collection and the operational management of the EI sector.

The remainder of this chapter draws from international experience to outline
a typical institutional structure for EI sector management, including the potential
division of roles and responsibilities across ministries and departments. This is a
general model that would need to be adapted to local circumstances. The type of
institutional structure that a country selects to manage the challenges of natural
resource extraction will depend to a significant extent on the country's political,
historical, and geological circumstances, as well as on the type of resource in
question.

Role of the Sector Ministry

The ministry of resources (oil, gas, and/or mining) is responsible for the overall
management of the EI sector, including setting resource policy, drafting laws and

regulations, and supervising EI sector agencies. Following are the typical functional components of an EI sector ministry:

- *A unit responsible for managing exploration and extraction rights*, which receives license applications and administers exploration and extraction rights, enforces license conditions, and maintains an up-to-date register of exploration and production licenses (for mining, this registry function is fulfilled by an up-to-date mineral rights cadastre).
- *An inspectorate*, which regulates the sector and monitors operations. Its responsibilities normally include developing technical specifications and standards; metering and monitoring production; technical supervision of operations; and ensuring compliance with licensing conditions, laws, and relevant regulations. It also monitors compliance with occupational health and safety standards. The monitoring of environmental standards may also fall to the inspectorate or be performed by the ministry of environment.
- *A geological survey*, which develops and maintains reliable information on national earth science infrastructure, including geological maps and related databases. It also provides the basic geological knowledge for the EI industry and relevant factors such as water resource management, environmental management, land use, geohazards risk management, and infrastructure works.
- *A mineral economics unit*, which analyzes economic issues in the EI sector. This unit should also analyze the economics of mining and hydrocarbon companies in the sector. It may be combined with the mining development/promotion unit to promote the sector at national and international events.
- *A promotion unit*, which promotes the sector among potential national and foreign investors. This could be combined with the mineral economics unit.
- *An artisanal and small-scale mining unit*, which addresses issues specific to small-scale mining in countries that have a significant artisanal mining sector.
- *A health and safety unit*, which establishes and enforces health and safety guidelines in the sector.
- There may be *a unit for monitoring the performance* and management of a state-owned enterprise (SOE).
- There may be *a policy unit* responsible for drafting and reviewing policies and laws.

These main functions will, in practice, be further disaggregated and complemented by other relevant functions, according to context. Figure 3.1 provides an example of an EI sector governance structure: a proposed organizational model for the Ministry of Mines of Afghanistan developed with support from the World Bank's Oil, Gas, and Mining Division (SEGOM).

Figure 3.1   Proposed Model for the Organization of Afghanistan's Ministry of Mines

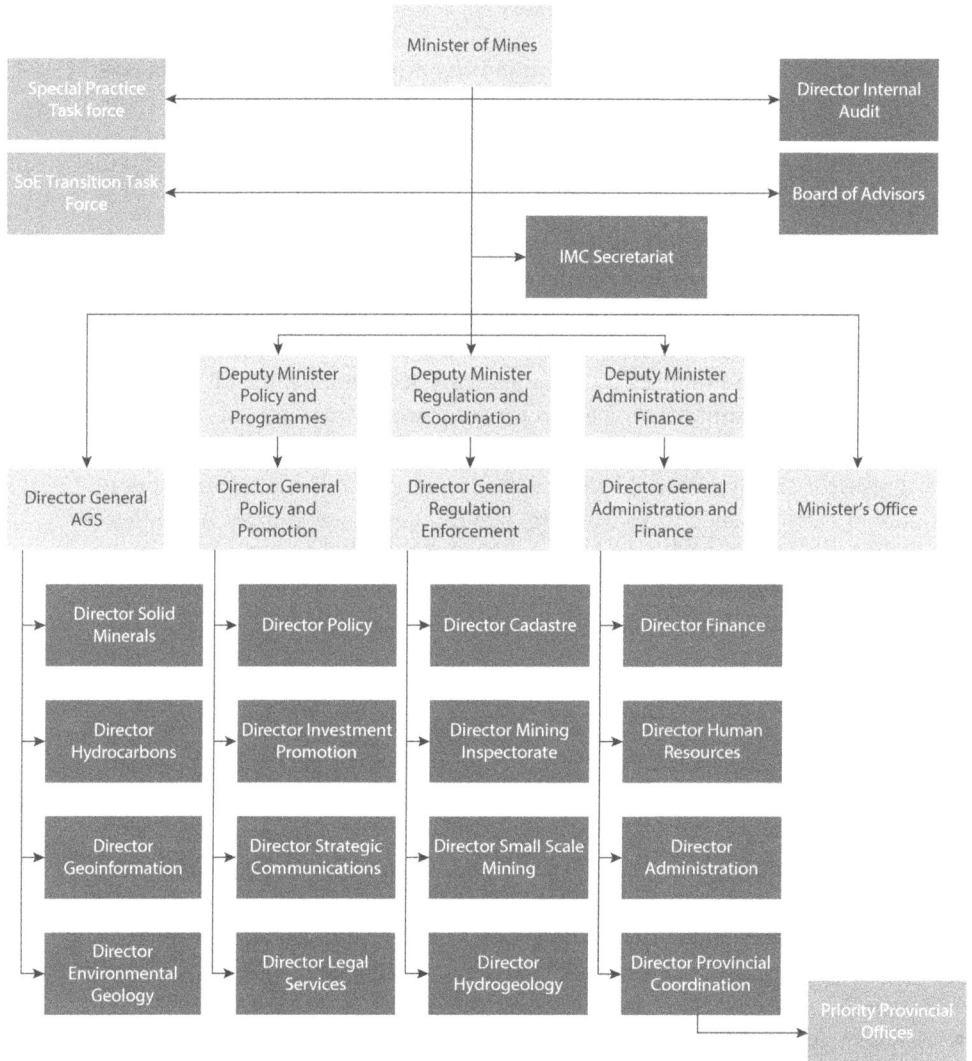

Source: Ministry of Mines and Petroleum, Government of Afghanistan (2010).
Note: AGS = Afghan Geological Survey; IMC = Inter-Ministerial Committee; SoE = state-owned enterprise.

Roles of the Ministry of Finance and Revenue-Collecting Agencies

The functions core to the implementation of public economic and financial policy typically lie with the ministry of finance (or equivalent ministry) and revenue-collecting agencies. These functions include several critical to EI sector management: (i) resource revenue policy (through the design of fiscal regimes), (ii) resource revenue forecasting and collection, (iii) the management of budgetary allocations to EI sector activities, and (iv) the management of liabilities that may arise from the EI sector.

In designing the EI fiscal regime, the ministry of finance requires the input of the sector ministry to ensure that policy encourages revenue collection but does not negatively affect the wider EI sector investment climate. As highlighted in chapter 5, in the section on EI fiscal regimes, these regimes may take multiple forms, with varying implications for revenue collection and also for promoting exploration and extraction. Technical input from the sector ministry regarding the global extractive investment environment is crucial when designing fiscal instruments and deciding appropriate tax rates.

The ministry of finance and the tax-collecting authority often have the primary responsibility for enforcing EI fiscal terms and collecting resource tax and nontax revenues, but with significant input from the sector ministry. Resource revenue is a function of several variables: the quantity and quality of the resource, the resource price, and the estimated production costs. Assessing these variables requires technical expertise in the extractive industry, to undertake physical audits that assess the quantity and quality of the resource; an understanding of the global resource market, to determine the appropriate resource price, particularly when commodities are not traded on exchanges; and knowledge of a mine's production and operational activities, to determine production costs and risks. The ministry of finance may not be best placed to have the requisite technical expertise; thus, it is important that there be strong and ongoing cooperation with the technical ministry in charge of oil or mining and a clear delineation of roles and responsibilities.

The efficient collection of taxes and royalties from the EI sector requires staff with sufficient and relevant qualifications, but this critical human resource is often absent in many resource-rich developing countries. Highly qualified technical staff is needed to put the government on equal footing with multinational companies and large investors, and to ensure appropriate revenues for the government and successful management of the EI sector in general. Given the large share of revenues that come from the EI sector in many resource-rich states, and the challenges linked to EI revenue management, it is appropriate that this sector should receive staffing time in some proportion to the share that it contributes to the public revenue. In some countries, this arrangement is achieved by EI sector tax authorities being part of a larger taxpayer unit within the revenue authority; others have a dedicated EI sector tax unit. Regardless of institutional form, the ministry or agency tasked with revenue collection must have highly skilled and experienced staff to implement EI revenue management efficiently, and be able to retain staff in the face of high salaries offered by the private sector.

Another key function for the ministry of finance, or another designated agency within the government, is forecasting resource revenues to feed into medium-term budget projections. Forecasting resource revenues often proves challenging on account of volatility in global commodity prices and difficulty estimating the potential production of EI companies operating in the country. Tracking these key production and price variables requires specialist expertise in the dynamics of global commodity markets (and, again, the input of the sector ministry). Production plans are critical to providing bottom-up estimates of production, and should be submitted by the companies to the sector ministry for monitoring purposes, and in turn shared with the agency responsible for producing revenue forecasts. Bottom-up production plans should be verified and monitored through physical audits conducted under the supervision of the sector ministry. Other relevant responsibilities for the ministry of finance to uphold include the following:

• Allocating necessary resources, through the budget process, to other ministries and agencies, to support the efficient and effective management of the EI sector
• In countries where revenue is shared at the subnational level, ensuring that resource revenues are shared with subnational governments as per the legislated (or regulated) revenue-sharing framework; readers interested in subnational sharing of revenue are referred to Anderson (2012), McLure (2003), Ahmad and Mottu (2003), and Brosio (2003)
• Participating in the formulation of resource policy and administrative guidelines, in particular with respect to revenue collection and administration, design of fiscal rules, and management of resource revenues (including the creation and governance of sovereign wealth funds, SWFs)
• Participating in transparency initiatives across government to ensure the transparency of resource contracts and resource revenue (in countries participating in the implementation of the Extractive Industries Transparency Initiative, this could be achieved through active participation in the multistakeholder group for this initiative)
• Monitoring the buildup of liabilities arising from extractive exploration and production, including from maintenance costs associated with local public infrastructure built by mining companies, costs of environmental damage, and costs of resettling displaced communities

Role of the National Resource Company

Interested readers are referred to McPherson, "State Participation in the Natural Resource Sector" in Daniel, Keen, and McPherson (2010), The Taxation of Petroleum and Minerals, Principles, Problems, and Practice, *which provides an excellent discussion of relevant issues, and on which this section draws.*

National resource companies (NRCs) have a large role in EI sector management. In the oil sector, national oil companies (NOCs) control 90 percent of world oil reserves and account for over 70 percent of production (BP Statistical Review 2008). In the mining sector, national mining companies have in many cases been replaced by state ownership through share participation, either on a free-carried or paid-in-equity basis.

The establishment of NRCs has been motivated by a range of commercial and noncommercial objectives. The chief objective often cited is to generate additional

revenues for the state, through dividends and taxes arising from commercial profits. NRCs have also frequently been called upon to fulfill a wide range of national, economic, social, and political noncommercial objectives, such as job creation
and provision of social and physical infrastructure. The noncommercial, quasi-
fiscal, and frequently off-budget activities of many NRCs have often precluded
efficient budget management and macroeconomic policy (Cameron and Stanley
2012). The complexity arising from such mixed objectives—many of them not
measurable (or measured)—has also contributed to a lack of transparency in
many NRCs.

With regard to their commercial objectives, and in the mining sector in particular, NRCs carry a level of risk that is not necessarily compensated by the
benefits, especially when compared against a well-designed and efficiently
administered fiscal regime. Government financing of loss-making NRCs can draw
funds away from other priorities, jeopardizing developmental objectives in education, health, and housing, among other areas. Simulations of fiscal regimes
suggest that revenue gains from dividends are small where modern, efficient tax
systems are applied (McPherson 2010, table 3).

Furthermore, in pursuing their commercial objectives, NRCs often assume
both regulatory and commercial responsibilities, creating conflicts of interest
that may be associated with poor commercial results. Such conflicts of interest include being simultaneously engaged as a partner to a private investor
and as a regulator of the same investor—or indeed as a self-regulator of one's
own commercial operations. International experience shows that the most
successful NRCs tend to be those whose regulation and noncommercial functions are assumed by the government, and that have limited noncommercial
objectives and are subject to competition from other companies (Heller,
Mahdavi, and Schreuder 2014). Norway's NOC, Statoil, is one such example.
From its establishment it has been exposed to international competition,
encouraged by a government expecting its efficiency to benefit from competition, risk sharing, technology transfer, and the influx of petroleum management skills (McPherson 2010).

Consistent with their independent commercial role, NRCs should in legal
terms be established as separate entities under corporate law, and not as units of
a government ministry. By the same principle, their management staff should be
appointed on the basis of professional qualifications, rather than by political
appointment processes, and boards of directors should ensure that companies
focus on their core business rather than the pursuance of political aims.

Reflecting these experiences and concerns, the main responsibilities of an
NRC include the following:

- Managing the commercial aspects of state participation in the EI sector
- Developing specialized expertise in the EI sector
- Optimizing value for shareholders (whether the state owns all or a portion of
  shares)

Conversely, responsibilities of the government (as represented by its EI sector ministry and regulatory agency) in overseeing the NRC may include the following:

- Maintaining focus on the state's commercial interests and, in line with this, potentially facilitating the NRC's expansion abroad
- Ensuring a competitive environment among EI companies, including the NRC
- Benchmarking the NRC's commercial performance against other companies
- Achieving an adequate distribution of responsibilities between the NRC and other government entities

Roles of Other Ministries and Government Agencies

In addition to the ministry of mines (oil and gas) and the ministry of finance, various other ministries and government entities are likely to be involved in the implementation of EI sector policies. Examples of these entities and their relevant responsibilities are listed as follows.[1]

- *The ministry responsible for justice and constitutional affairs:*
  - Drafts the oil, gas, and mining legislation, with technical input from the sector ministry and the ministry of finance (for fiscal issues)
  - Drafts legislation for resource revenue management, with technical input from the ministry of finance
  - May draft regulations relating to mineral licensing and regulation of mining operations

- *The ministry responsible for the environment, water, forests, wetlands, and wildlife:*
  - Ensures that an appropriate closure plan is developed in the early stages of project development (see chapter 6 on environmental safeguards)
  - Monitors compliance with environmental protection plans and closure plans, and levies fines or other penalties permissible by regulations for non-compliance
  - Regulates the use of water and pollution through the issuance of water and pollution permits
  - Ensures compliance with conditions of water and pollution permits, as well as protection of water catchment and drainage areas, forests, and wetlands
  - Monitors the impact of resource extraction on the quality of ground- and surface-water bodies, soil, flora, and air in the area of the extractive activity
  - Ensures that EI sector policies are consistent with wildlife conservation
  - Manages any potential impact of toxins arising from activities in the EI sector

It may be noted here that best practice necessitates that environmental issues be addressed by specialized ministries—primarily the ministry of environment or its equivalent—rather than the EI sector ministry. The EI sector ministry may nevertheless retain a small environmental unit to coordinate with the specialized ministries in charge of environmental and social issues.

- *The ministry responsible for industry:*
  - Ensures that resource policies are coherent with national policies for industrial development
  - Promotes development of upstream, downstream, and sidestream production linkages to qualified domestic companies, in conjunction with the sector ministry
  - Facilitates, in cooperation with the private sector, training for firms and workers that need skill upgrades to establish production in the upstream, sidestream, or downstream sectors

- *The ministry (or ministries) responsible for education and research:*
  - Establishes and/or promotes education and training programs that address specialist staffing needs in the extractive and related sectors (such as oil and mine technicians, engineers, geologists, and so on), with input on design provided by the sector ministry
  - Provides support to the sector ministry in facilitating technology transfer by means of extractive-related research and development (R&D) activities in national universities and research institutions, jointly with multinational resource companies and other relevant institutions

- *The ministry responsible for physical planning:*
  - Elaborates area plans for extraction locations, with input from the sector ministry
  - Coordinates oil, gas, and mining infrastructure development with regional and national infrastructure priorities, plans, and projects
  - Supports the ministry of transport and the sector ministry to evaluate options for the establishment of resource corridors and industrial clusters

- *The ministry responsible for labor:*
  - Ensures that employment policies in the EI sector are consistent with national employment policies and regulations
  - Monitors compensation practices for occupational injury and illness
  - Mediates labor disputes and participates in conflict resolution

- *The ministry responsible for local governments:*
  - Contributes to and coordinates local government policies for the extractive sector in accordance with the national policy for local development
  - May audit community development agreements, foundations, trusts, and funds
  - Provides guidance to local governments on planning and capacity building related to activities in the EI sector
  - Integrates extractive activities in the local government's plans and programs
  - Ensures that oil, gas, and mining infrastructure development is integrated into local development plans, if relevant through shared-use agreements

In addition to the above-mentioned entities, a number of other ministries and government agencies are involved, as follows:

- *The central bank* addresses the monetary and exchange rate effects of the EI sector; may play a role in the tracking, reporting, and reconciling of EI fiscal and financial flows; and, if relevant, may manage the saving and/or stabilization fund.
- *The auditor general* provides independent oversight of government operations in the extractive sector through financial and other audits (in accordance with constitutional provisions and other relevant legislation), and ensures adherence to international accounting standards in the oil, gas, and/or mining sectors.
- *The ministry responsible for information and communication technology* formulates regulation and legislation that addresses the provision of data transmission and storage in response to increased demand from resource companies for the same.
- *The ministry responsible for public works and transport* leads the planning and regulation of EI-related transport services, the development of EI-related infrastructure, and if relevant takes the lead in the planning of resource corridors.
- *The ministry(ies) responsible for security*, in contexts where security is a concern, responds to demands for security of oil, gas, and mining installations. This may be the case where the local population increases dramatically with the arrival of migrant workers to the mine.

Note

1. This section draws on Government of Uganda (2008).

CHAPTER 4

Investment and Production Cycles

Characteristics of Extractive Industry Investments

Investment in the extractive sector differs fundamentally, and in various respects, from investment in most other sectors. First, initial investments in infrastructure and mine/oilfield development can be very large; capital costs often run into the billions of dollars. Second, extractive industries (EI) investments are characterized by long lead times, since mines and oilfields can take a decade or more to discover and then develop. Once the expenditure is made, these investments are essentially sunk costs and must be completed before the mine or oil/gas field is brought into production. On the other hand, investment horizons can also be very long, with some mines producing for 50–100 years or more. Third, investments are subject to a high level of uncertainty, particularly at the exploration, design, and development stages, but risk remains high during operation due to oil and mineral price volatility, and uncertainty about the extent of reserves. At low prices, low ore grades are unprofitable to mine and are considered as waste. Similarly, oil and gas extraction that requires expensive technology is only profitable as long as the price remains high enough to cover the requisite expense. Fourth, oil and minerals are finite resources, and since production of an additional unit today means that this unit will not be available in the future, optimal depletion strategies are affected and differ depending on private and social time preferences. Finally, in the mining sector the costs of closing a mine can constitute a significant share of a project's overall budget.

The Mining Cycle

The mining cycle can be divided into four stages: exploration, design and construction, operation, and closure and postclosure. See figure 4.1 and table 4.1 for illustrations of this point.

Mineral exploration is the first phase of the mining cycle. In areas where minerals or metals have not been found before, the success rate is extremely low. If

Figure 4.1   The Four Stages of the Mining Cycle

Source: Authors' own compilation based on Newmont (2013) and Minerals Council of Australia (2015).

so-called grassroots exploration leads to a discovery, further examination is undertaken to decide whether the discovery can be taken to the next stage. On average only an estimated 1 in 10,000 discoveries leads to the development of a mine (Government of Canada 2006). The success rate at the exploration stage can be greatly increased by the use of geological maps and data of the kind produced by geological survey entities. A minimum level of geological information is required to reduce risks to a level acceptable for the private sector to begin exploration. (Where basic geological information is absent, risks are generally too high to trigger private sector investment.) From an economist's perspective, two market failures create space for the state to play a role. These occur when (i) early-stage geoscience has public good aspects, which means it will be undersupplied by the private sector; and (ii) absent basic geodata, the uncertainty associated with a successful discovery prospect is so high, and its potential value is so speculative, that investors cannot value it properly.

Once a mining claim is staked and approved by the relevant government agency, the prospector or company holds the exclusive rights to explore that piece of ground for a determined amount of time. Surface fees and obligations to invest are used to discourage companies from holding on to land for speculative purposes without undertaking further investment. Exploration may in some regions be undertaken by very small, "junior" mining companies that sell the extraction rights to larger companies if finds are made. Exploration is a very slow process: the time from the discovery of a promising mineral to the start of a new mine is usually 7–10 years, sometimes longer.

Mine development is the second phase of the mining cycle. The purpose of this phase is to estimate the potential value of the mineral deposit and determine if it can be mined profitably. Factors that determine profitability include the following:

- Location of the resource, type of mineral, and infrastructure requirements (for example, it may be possible to mine diamonds in a remote area with little infrastructure due to their high value-to-volume ratio—in contrast with lower-value and higher-volume minerals such as zinc or lead)
- Size and value of the resource
- Market prices and distance from markets
- Ability to mine the mineral in an environmentally safe manner
- Regulatory regime
- Availability of a qualified workforce

Table 4.1 The Mining Cycle

	Stage	Activity	Government approval	Cost estimate ($)	Typical duration
Mineral exploration	Geological infrastructure	Regional airborne geophysics, geochemistry, and geological surveys		10–100/km²	10–20 years, national program cycles
	Mineral resources assessment	Regional data integration, ore deposit modeling, prospectivity assessment			10–20 years, national program cycles
	Reconnaissance	Semi-regional airborne and ground surveys to identify targets		0.5–2 million	1–3 years
	Exploration	Trenching, ground geophysics, detailed geochemistry and geology, drilling	Exploration license; Exploration ESIA	0.1–50 million	1–10 years
	Advanced exploration	Drilling, pilot tests, prefeasibility study			
Mine development	Mine planning	Feasibility and engineering studies, ESIA and ESMP	Mining license; ESIA and others (for example, water)	0.5–10 million	1–3 years
	Construction	Infrastructure, mine development, processing facilities		50 million–15 billion	1–5 years
Mine operation	Mining	Ore production (open pit, underground, or alluvial)			10–100 years
Mine closure	Closure	Final closure and decommissioning	Release of license	1–50 million	1–5 years
	Postclosure	Maintenance		0.1–0.5 million/year	Perpetuity

Source: Adapted from Ortega Girones, Pugachevsky, and Walser (2009).
Note: ESMP = environmental and social and management plan; ESIA = environmental and social impact assessment; km = kilometers.

The main activities of the mine development stage include the following:

- Acquiring additional technical, environmental, and socioeconomic data
- Developing the mine, infrastructure, and closure plans
- Ensuring that regulation requirements are satisfied (through, for example, consultations between mining companies and the government)

- Evaluating environmental and socioeconomic effects
- Obtaining permits and licenses and setting up the closure surety mechanism (for closure surety, see chapter 6)

Mine design and construction usually takes from 5 to 10 years, depending on the location of the mine, its size, and the complexity of the development process, including infrastructure needs. At the end of the mine development stage, a final evaluation of the project is completed and a decision is taken on whether to produce or not. If the decision is positive, the mine and its facilities are built. In general, it takes:

- Two to three years for studies and tests (feasibility and associated environmental studies, table 4.2)
- One to three years for permitting
- Two to four years to build the mine and infrastructure

The costs of mine development can vary from around $50 million to $15 billion, according to the mine type and size, location, and infrastructure requirements.

Mine operation is the third phase of the mining cycle. The operating life of a mine can be as short as several years, but in the case of major operations it can stretch over several decades—50–100 years or more. An operating mine has four main work areas, for excavation, the processing plant, waste storage, and supporting services. Some mine operations do not have a processing facility on site, and the ore is sent elsewhere for processing. Factors that affect the operating life of a mine include the following:

- The mineral price on the world market
- Production costs and production rates
- The quality (grade) of the extracted ore
- The size and shape of the ore deposit, as well as its depth below the surface
- Mining methods and equipment, and related costs
- Safety issues related to ground conditions

During mine operations, the highest cost is usually of labor, including training. Fuel, power, and other consumables (heavy equipment, drill bits, spare parts, explosives, chemical reagents, and so on) are other important expenses.

Mine closure, the fourth state of the mining cycle, is defined as the orderly, safe, and environmentally sound conversion of an operating mine to a closed state. The objective of mine closure is to leave areas affected by mining activity as viable and self-sustaining ecosystems compatible with a healthy environment and with human activities (Government of Canada 2006). Mine closure also has significant social implications, linked to the loss of employment and income in the local community.

Table 4.2 Feasibility Studies: An Overview

Geology and resource determination	*How large is the deposit or resource?*
	What is the grade of the minerals or metals in the deposit?
Mine planning	What will be mined? (what metals, how much of the deposit?) How will it be mined? (open pit/surface or underground?) What equipment will be used to mine it?
Process plant test work and plant design	What is the best way to extract the minerals or metals? (for example, from the host rock?) Which minerals or metals will be recovered? What is the estimated recovery level of the chosen method? Will there be a smelter or refinery?
Infrastructure planning	What roads, airstrips, camps, and complexes will be needed?
Water and waste management planning	What are the water supply needs? What are the discharge quality requirements? How can waste be safely disposed of?
Environmental and socioeconomic planning	What are the main concerns raised by environmental and socioeconomic studies? How can plans address these concerns?
Mining closure and reclamation plan	What are the best approaches to reclamation and closure?
Operating cost estimates	How many workers are required? What types and amounts of equipment and supplies are required during operations? What are the annual operating costs?
Capital costs	What are the costs to plan, design, permit, and construct the facilities?
Financial analysis	What are the potential financial returns to the mining company? What are the costs for borrowing money to build and operate the mine? What are the yearly costs and annual earnings? What is the expected profit or loss?

Source: Adapted from Natural Resources Canada (2013). Reproduced with the permission of the Department of Natural Resources, 2015.

Planning for mine closure starts during the planning phase of the mining project. In most countries, the government must approve the initial reclamation plan prior to mine development. The company involved must also put up a closure surety (deposit, bond) to ensure that funding for reclamation is available even if the company were to go bankrupt or renege on its responsibilities. The closure surety ensures that governments are not left with liabilities following incomplete mine closures.

A mine closure and reclamation plan details how the mining company will close the mine site and, as far as possible, return it to its premining state. The plan sets out what to do with each individual component of the mine. The time needed for mine closure depends on the size and complexity of the operation and on the environmental effects of the mine. Typically, it takes 2–10 years to shut down a mining operation, but if long-term monitoring is required, then the closure may take decades to complete. The cost may range from $1 million to $50 million or more, with postclosure monitoring and maintenance costing from $0.1 million to $0.5 million per year. In the case of permanent environmental issues, postclosure monitoring and maintenance may last indefinitely. Chapter 6 describes the financial mechanisms used to ensure that mining companies complete the closure and reclamation stage appropriately, thereby avoiding potentially large liabilities for the government.

The Oil and Gas Cycle

The stages of oil and gas operations are similar to those of mining:

- *Exploration surveys and drilling.* Geological maps are reviewed in desk studies with the objective of identifying major sedimentary basins. Once a promising geological structure has been identified, the only way to confirm the presence of hydrocarbons is by drilling exploration wells, either from a pad constructed for the purpose (on land) or from an offshore drilling unit (at sea). Where a hydrocarbon formation is found, initial well tests (which may take another month) are undertaken. Unlike minerals, exploration for hydrocarbons (including coal bed methane) requires pilot-level (test) production to confirm feasibility. This is because reservoir porosity and interwell connectivity (during production) are equally important determinants of potential, as are resource size and quality.
- *Appraisal.* When exploratory drilling is successful, more wells are drilled to establish the size and extent of the field, using the same techniques as for exploratory drilling.
- *Development and production.* Once the size of the oilfield is established, production wells are drilled. Large reservoirs require various wells to be drilled. Advances in directional drilling have increased the number of wells that can be drilled from one location to as many as 40 or more. After reaching the surface, the hydrocarbon is routed to a central production facility, which gathers the produced fluids (oil, gas, and water) before they are transported onward for further processing. The typical commercial life of an oilfield, although there is ample variation across fields, tends to be in the range of 10 to 30 years.
- *Decommissioning and rehabilitation.* As with mining, the decommissioning process should be considered at the initial planning stage of the operation. The site should be restored to environmentally sound conditions. The local environmental footprint of oil extraction is much smaller than that of mining; for land-based production, it is essentially just a concrete production pad and adjacent infrastructure, wells that need to be plugged, and pipeline infrastructure. The cost of decommissioning is therefore significantly lower for hydrocarbons than for minerals. Exploration wells, most of which will be unsuccessful, are normally decommissioned after one to three months of activity.

CHAPTER 5

Extractive Industries Policy

Policy and Regulatory Frameworks

For an excellent overview of petroleum sector management, see Tordo, Johnston, and Johnston (2010), "Petroleum Exploration and Production Rights." For a private sector perspective on the role of mining in social and economic development, readers are referred to Kapstein and Kim (2011), "The Socio-Economic Impact of Newmont Ghana Gold Limited." See also ICMM (2013), "Approaches to Understanding Development Outcomes from Mining."

To contribute optimally to economic development and growth, the extractive industries (EI) sector requires stable and predictable public policy frameworks. Countries differ in which strategies and policies they choose to realize their objectives for the extractive sector. Deciding factors include the current structure of the country's public administration, geology, and history, as well as the amount or type of investments involved. In all cases, a certain set of functions needs to be fulfilled, independent of external circumstances. To attract investment, geological data must be made available (to provide indicators of the size and quality of potential deposits), and a favorable investment climate needs to be established. Equally important, a legal and cadastral regime must be put in place to manage exploration and extraction rights. Environmental and social safeguards need to be included from an early stage of project development. A stable and transparent legal framework, regulations, and a licensing regime must be established, and an appropriate administrative framework set up for revenue collection. This chapter provides an overview of the main components of extractive policy regimes.[1]

A mineral policy sets the framework for mining sector objectives, as defined in a democratic and consultative process involving the government, civil society (including potentially affected communities), and the private sector (box 5.1). Components include (i) defining the mechanisms by which the government will create a competitive business environment, (ii) establishing provisions to attract sector participation, (iii) finding a balance between direct taxation (through taxes, royalties, and surface fees) and indirect taxation of secondary

---

Box 5.1 Mineral Policy

A well-prepared mineral policy should aim to accomplish the following:

- Establish guiding principles for current and future generations to use resources.
- Create an enabling environment for local and international investments in mineral exploration, development, and production.
- Encourage governments, companies, and communities to work together to ensure that operations are conducted in an environmentally and socially sustainable manner by (i) avoiding, mitigating, or compensating for local impacts; (ii) facilitating effective community consultations; (iii) mandating effective community development plans; and (iv) using sustainability indicators to track and report outcomes.
- Ensure local economic benefits, revenue sharing, and the transparent management of revenues.
- Help make informed decisions through the collection of geological data and resource assessment.
- Regulate artisanal and small-scale mining and provide for adequate institutional and technical support for related activities.
- Apply internationally accepted standards for environmental and social protection, including for indigenous peoples and the resettlement of affected communities.

Source: Adapted from Stanley and Eftimie (2005).

---

and tertiary support industries and downstream processing, (iv) introducing revenue transparency initiatives, and (v) ensuring the use of concessions, mineral leases, and other assignable mineral rights for essential commodities (Stanley and Eftimie 2005).

International experience demonstrates the need to clearly separate policy, regulation, and commercial activities (table 5.1). By this arrangement a government ministry directs policy, a regulatory body provides oversight and expertise, and a national resource company (NRC), if there is one, engages in commercial operations. Such a separation of powers sets up a system of checks and balances, ensuring greater systemic integrity and avoiding the conflicts of interest that can arise when commercial, policy, and regulatory functions are consolidated in fewer entities. If regulatory functions cannot be separated from the NRC, another option is to ring-fence these functions for operational and accounting purposes and to provide direct reporting to the national budget and accounts. This offers a temporary solution until the necessary capacity and credibility can be established in an external agency.

Financial resources are needed to implement institutional and regulatory frameworks associated with EI policies, including adequate and sufficient training for relevant staff. Oil and mineral policy makers need to recognize that monitoring and implementing mining-, oil-, and gas-related activities introduce

Table 5.1 Separation of Key Functions in the Extractive Sector

Policy	
Finance Ministry (fiscal)	Sector Ministry
	(resource management)
Regulation	
Finance Ministry	Sector Ministry
(tax office)	
Commercial	
Natural Resource Company	

Source: Calder 2010. Copyright 2010 Routledge/IMF; reproduced by permission of Taylor & Francis Books, UK.

additional responsibilities not only for the sector ministry and the ministry of finance but also for other ministries and agencies (as described in chapter 3). The ministry of finance (or its equivalent) will in this context have to evaluate the need for additional resources to carry out these responsibilities and to provide funding as considered appropriate. Table 5.2 summarizes the requisite components of an EI sector program.

Sector Financing, Ownership, and Liabilities

This section summarizes and condenses material from Cameron and Stanley (2012), EI Source Book, chapters 6 and 7. See also the World Bank (2011a), "Overview of State Ownership in the Global Minerals Industry."

In many countries, legislation allows direct state participation through minority equity holdings in EI ventures. The equity can be held directly in the name of the government, or it can be held by a government vehicle established to hold equity in private companies. Whereas state participation was usual in both the mining and the oil sectors until the 1960s and 1970s, state participation is currently far more common in the oil sector. Several types of state participation can be observed in the EI sector: (i) paid-in capital participation, (ii) carried interest participation, and (iii) free equity participation (see box 5.2). In addition, production sharing is common in the oil sector (discussed in the section titled "Overview of Extractive Industries Tax and Royalty Regimes"; see figure 5.3 on production sharing).

Minority equity participation provides the government with shareholder information about the project and the private partner company—information that is not necessarily otherwise available. Whereas equity participation can be motivated by nonfiscal objectives, it typically also carries an expectation of profit through a share in dividends.

Cameron and Stanley (2012, section 7.1) argue that minority equity participation adds little to government revenue compared with what could be

Table 5.2 The Components of an Extractive Industries Sector Program

Sector financing, ownership, and liabilities	Determines the type of state participation in extractive ventures, if any; options include (i) paid-in capital participation, (ii) carried interest participation, and (iii) free equity participation. Includes shareholder agreements that address the (i) type of funding of the equity, (ii) decision-making powers of various shareholders, (iii) conditions under which ownership may change, and (iv) shareholders' responsibilities and obligations at the time of decommissioning/closure.
Legal, regulatory, and licensing framework	Establishes the framework for the sector, including ownership of minerals and assignation of mineral rights; the role of the state; compliance with mineral rights laws; organizational structure of the sector; and terms, conditions, and procedures for project approval.
Geodata information system	Makes geodata available to potential investors and the government by establishing and maintaining a user-friendly geodata database available to the public, undertaking new geological surveys, and developing geosciences laboratory capacity to undertake reliable analysis of explored tracts.
Mineral rights cadastre	Ensures transparency in the granting of mineral rights, guarantees security of tenure, and facilitates the management of competing land uses. Compiles mining titles and their history and specific attributes in a common database, freely available to the general public via the Internet.
Mineral tax regime	Determines the composition and magnitude of tax instruments, including the options of corporate profit tax, royalties, progressive tax instruments, cost-recovery provisions, and ring-fencing.
Tax administration	Establishes the capacity to manage complex tax operations associated with extractive industries. Includes establishing the requisite specialist staff capacity or obtaining external specialist support, and setting up appropriate organizational structures (such as a large taxpayer office), procedures, and information technology (IT) infrastructure.
Contract management capacity and transparency	Provides for internal (government) oversight and external (society) accountability through the regular release of comprehensive information on resource licenses, contracts, and revenues. Integrates transparency considerations in administrative procedures for contract and revenue management.
Environmental objectives and potential liabilities, mine reclamation	Avoids or reduces environmental liabilities that are potentially costly to the government by enforcing environmental regulations, including the elaboration and implementation of environmental impact assessments, environmental management plans, and closure/decommissioning plans.
Social objectives and potential liabilities, health and safety	Ensures that stakeholder consultation is implemented among local communities and provides a legal basis and incentives for community development agreements and community foundations, trusts, and funds. Prepares for the economic and social viability of communities after the mine is decommissioned.
Extractive industries business climate/ competitiveness	Reduces risks and costs of doing business for resource companies, including through secure property rights, predictable regulation, and a light administrative burden (by reducing red tape).
Public investment of resource revenues	Ensures that subsoil resources are sustainably invested in human capital, infrastructure, and, generally, in increased non-resource-based productivity.

Source: World Bank.

accomplished by an efficient, flexible, and well-administered tax regime—but may add considerable risk. On the revenue side, dividends (and associated revenues like withholding taxes on dividends) are unpredictable and may take years to appear because of high initial investments and volatile commodity prices. Regarding risk, the government may, as a minority shareholder, have limited

---

Box 5.2   Modes of State Participation

Paid-in-capital participation. In this mode, the government pays for equity in cash (or through the contribution of a license or other assets), giving it the same standing as other shareholders. Government investment decisions are often made as part of the overall budget process and, ideally, according to potential profits. (This may not be the case where the state-owned enterprise is partially listed or makes independent capital allocation decisions.)

Carried interest participation. This type of participation may take several forms, the most frequent of which is the so-called partial carry. Under this approach, the private investor "carries" or pays the way for the government-owned partner (the national resource company, NRC) through the early stages of a project—the exploration, appraisal, and (possibly) development phases. After some predefined milestone, the government (that is, the NRC) spends an amount proportionate to that of the private investor, as under the paid-in-capital participation model. The advantage for the government is that it does not have to provide cash up front. For the company, the carried interest dilutes its equity base, since it must raise the cash needed to cover the government's participation. Carried interest is thus essentially an equity contribution from the shareholders on behalf of the government. A "full carry" occurs when all costs are borne directly by the private investor, and compensation is paid out of the state's share.

Free equity participation. In some instances, governments insist on a minority free equity in a new extractive project. Free equity is tantamount to taxing the project (that is, when the project is generating dividends, it is effectively the same as withholding tax on dividends), but unlike taxation it also gives the government the many obligations and risks of a shareholder. The mandatory requirements of free equity have reportedly created resentment and distrust between governments and companies.

Source: Based on Cameron and Stanley (2012).

---

decision-making power in contexts where all major decisions are taken by the majority shareholder. For example, if the company decides to raise additional equity from its shareholders, the government may face the choice between having to buy more equity or see its ownership share diluted. If the company is losing money, the government may be required to contribute additional funding to keep the company operating, and as a shareholder the government is also exposed to contingent liabilities associated with closure/decommissioning. Finally, conflicts of interest may arise if the government's regulatory role and its ownership role are not sufficiently separate.

Decisions on the government's minority equity role in EI sector projects should be based on shareholder agreements. These should address the (i) type of funding, (ii) dividend declaration rules, (iii) decision-making powers of different shareholders, (iv) conditions under which ownership may change, (v) materiality thresholds and rules for approval of material-related party transactions, and (vi) shareholders' responsibilities and obligations at the time of decommissioning/closure. Funding decisions should consider both initial capital expenditures and subsequent needs arising from cash-flow shortages or from funding that sustains capital expenditures or production expansions. The decision-making powers of the various shareholders include over dividends, budget approvals, senior management appointments and remuneration, investment programs, and the raising of new capital (including debt).

Mineral Legislation, Regulation, and Contracting Regimes

Interested readers are referred to the World Bank's Extractives Industries Contract Monitoring Roadmap; *Smith and Rosenblum (2011),* Enforcing the Rules; *and Rosenblum and Maples (2009),* Contracts Confidential: Ending Secret Deals in the Extractives Industries. *The Web site www.resourcecontracts.org contains a database of searchable oil and mineral contracts, as well as downloadable guides to understanding oil and mineral contracts. At the time of writing, the African Mining Legislation Atlas, www.a-mla.org, was being developed. For an explanation of oil contracts, see Open Oil's "Oil Contracts: How to Read and Understand Them." Commercial databases of mining regulations are also available.*

Mining and petroleum legislation establishes a framework for EI development, consistent with constitutional prerequisites and long-range national objectives (box 5.3). Based on the mineral policy, the mineral and petroleum laws define the rights, privileges, and obligations of those involved in the development of oil, gas, and mining projects. Mining regulations give a working application to each article in the law, describing the procedures and processes used in the

---

Box 5.3   Mining Law

A well-prepared mining law accomplishes the following:

- Establishes the ownership of minerals and assigns mineral rights
- Defines the role of the state
- Defines the characteristics of extractive industries' administration, including of the mines cadastre, inspectorate, and authorizations
- Ensures compliance with mineral rights laws, including royalty, surface fees, and other financial obligations
- Defines relations among mineral holders
- Establishes the basis for health and safety regulations, environmental protection, and measures to mitigate social impacts
- Ensures protection of investments, as well as mechanisms for dispute resolution and transitional provisions

Source: Stanley and Eftimie 2005.

---

administration of the sector. Below the mineral policy level, there is a chain of implementation that increases in detail and specificity as it passes from mining law and mining regulations, down to mining contracts and licenses.

Mining law is implemented through mining regulations. These define the procedures by which mineral rights can be acquired, transferred, expanded, rescinded, or otherwise modified; the administrative structure of the ministry responsible for the sector; and the role of each unit in applying the law (Stanley and Eftimie 2005). Mining regulations specify terms and conditions for project approval (licensing), for decisions regarding environmental and social impacts, and for monitoring of indicators and reporting of outcomes. Although the mining law specifies an EI fiscal regime, the regulations are what define this regime's exact mode of calculation, payment, and other procedures, and what provide requirements for environmental and social impact assessment and management plans (box 5.4).

Resource-rich countries need to be able to correctly implement licensing rounds, public offerings (tender/auction), and competitive and transparent tender packages. Exploration and extraction rights are normally granted by the government by means of concessions, leases, licenses, and agreements (box 5.5). Efficient and effective granting procedures tend to be based on the following principles:

- A clear legal and regulatory framework
- Well-defined institutional responsibilities
- Transparent and nondiscretionary procedures

The award of oil, gas, and mineral exploration and extraction rights follows one of three principal processes, depending on the type of resources, the exploration

---

Box 5.4 Mining Regulations

Well-prepared mining regulations accomplish the following:

- Define the organizational structure of the sector
- Establish terms and conditions, procedures for project approval (including environmental and social impact assessments), and the development of environmental and social management plans
- Define procedures for filing reports and forms
- Establish procedures for annual assessment of production
- Provide information on surface fees and charges for filing applications and transferring titles
- Establish standards for annual reporting

Source: Stanley and Eftimie 2005.

Box 5.5 Mining Contracts and Licenses

Mining contracts and licenses most often accomplish the following:

- Establish work commitments for exploration and development.
- Define economic and financial provisions specific to the project and in accordance with (i) leases and other rights for mineral development, (ii) provision of infrastructure, (iii) project financing, and (iv) currency and exchange controls.
- Establish specific parameters and processes for environmental and social management.
- Provide for local economic development.
- Ensure provisions for termination and settlement of disputes.

Source: Stanley and Eftimie 2005.

---

risk, and (potential) investor interest. Principles of efficient and effective mineral-rights allocation policies include (i) transparent, competitive, and nondiscretionary procedures for the award of exploration/exploitation rights; (ii) a consideration of ancillary infrastructure required to service development and production; (iii) the existence of clear legal, regulatory, and licensing frameworks; (iv) well-defined institutional responsibilities; and (v) clearly specified socioenvironmental safeguards. For geological reasons, coal and other resources where geological continuity can be well anticipated tend to be associated with low levels of resource risk.

When tendering and licensing rights for oil, gas, and mineral exploration and extraction, many developing countries find themselves at an informational disadvantage compared with multinational resource companies. Whereas companies have access to highly qualified international specialists in the areas of geology and mineralogy, law, finance, and other relevant disciplines, developing countries often lack that capacity. Such imbalances put governments in an unfavorable position when negotiating exploration and extraction contracts and licenses. It is fundamental to build capacity to manage EI contract negotiations. Developing the needed policy, fiscal, legal, and regulatory reforms and frameworks reduces the risk of politically difficult remediation or contract renegotiation at a later stage. The World Bank's Extractive Industries Technical Advisory Facility (EI-TAF) provides countries with assistance in negotiating oil, gas, and mining contracts (see table 5.3).

Table 5.3 Types and Characteristics of Mineral Rights Awards

Award type	Characteristics	Exploration	Exploitation
Direct negotiations	Least transparent process	Government provides direct exploration rights to investors, in exchange for information being generated.	Government negotiates with company for rights to a known resource, in exchange for other (infrastructure) investments.
Open mineral access	Assignment of mineral rights on a first-come, first-served basis	Award of exploration rights based on standard cadastral units to ensure they do not conflict with restricted areas or other exclusive mineral rights.	Holders of exploration rights have the security of tenure and are granted the right to transfer to mineral exploitation, subject to full regulatory compliance.
Competitive resource tenders	Assignment of mineral rights by competitive tenders	Exploration mineral commodities for which (i) geological information is already public knowledge, and (ii) more than one investor is potentially interested.	Given competitive exploration rights, the conversion to exploitation rights are subject to full regulatory compliance.

Source: Stanley and Mikhaylova 2011.

Establishing and Maintaining a Geodata Information Base

This section condenses material from BGS International (2012), "Geodata for Development, A Practical Approach." Interested readers are referred to the original document.

The effective acquisition, management, and dissemination of geodata are key to facilitating investment in mining, as well as in the oil and gas sector. Additionally, a government with a good understanding of the country's geological potential can use this informational base to better manage the country's subsoil resources. The high returns on the collection, storage, and public availability of geodata are linked to the following:

- Attracting investors, who will be able to use reliable geoscientific datasets to guide decisions and reduce risk
- Improving government capacity to negotiate contracts
- Facilitating additional oil, gas, and mineral discoveries
- Developing resource corridors for the regional coordination of mine site development with infrastructure development

Core outputs include the following:

- A user-friendly database of existing geological information, in which documents are cataloged and scanned
- New geological surveys of mineral and hydrocarbon tracts, used to identify prospective areas for exploration
- A modern geosciences laboratory, to conduct reliable analysis of explored tracts
- The development and maintenance of policies to retain and increase staff capacity

Figure 5.1   Sharing Costs of Geodata between the Private and Public Sectors

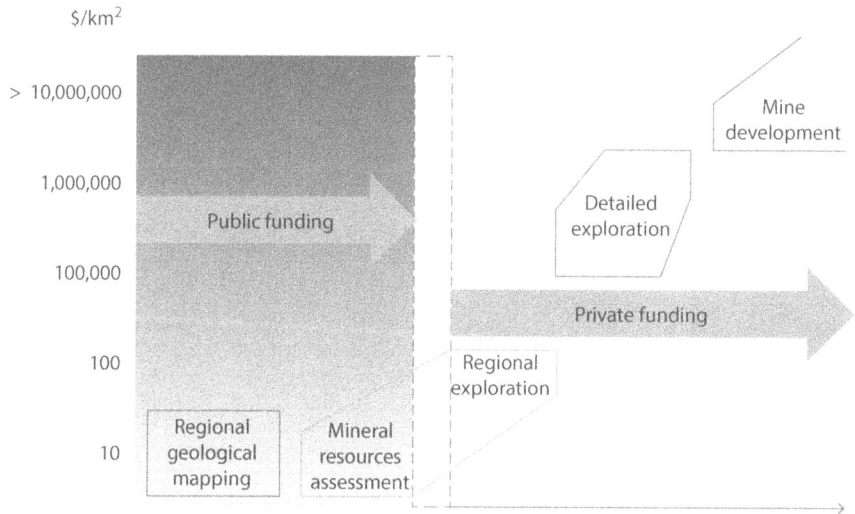

Source: World Bank Oil, Gas, and Mining Policy Division.
Note: km² = square kilometers.

The cost of acquiring and processing geodata information is normally shared by the public and private sectors (figure 5.1). In general terms, governments tend to provide information at the aggregate level (macrolevel data, regional geological mapping) to attract the initial interest of investors and interested companies. Companies then develop more detailed surveys at their own cost. The cost tends to increase with the level of detail, as the target exploration area narrows. The point at which the responsibility for geodata acquisition and processing passes to the private sector varies by country and by commodity.

When considering the cost of establishing geodata coverage, it is convenient to separate the cost of establishing the requisite institutional infrastructure from that of undertaking the concrete surveys. In resource-rich developing countries, donor-supported geological programs frequently address the building of essential infrastructure and local capacity, such as vehicles, laboratories, computers, and training. However, it has in several cases proved challenging to maintain the capacity of geological survey organizations built up with donor assistance. Frequently, staff members who have increased their value on the job market by receiving training at the national geological survey leave for better-paid jobs in the private sector, and donated equipment falls into disrepair. With databases no longer properly maintained, even previously acquired geodata information may no longer be publicly available, and there may be risks of data being lost.

The economic benefits of publicly available geodata can range from tens to thousands of times the cost (Reedman and others 2008). There is thus a strong case for funding geological surveys appropriately. But where this for some reason is not feasible, for-profit or subsidized, privately based options, or public–private partnerships

(PPPs), may be solutions. In any case, a business model needs to be developed that covers the long-term costs of maintaining and disseminating the data.

Mineral Rights Cadastre

This section condenses material from Ortega Girones, Pugachevsky, and Walser (2009), "Mineral Rights Cadastre." Interested readers are referred to the original document.

The establishment of a public register and the application of nondiscretionary, consistent procedures as part of the mineral rights cadastre are critical. These components ensure transparency in the granting of mineral rights, guarantee the security of tenure, and facilitate the management of competing land uses (for example, in the case of mining interests versus protected areas). Essentially, the mineral rights cadastre manages the mining titles in a country. When well developed and complemented by capable public mining sector institutions, the cadastre integrates regulatory, institutional, and technological aspects of mineral rights administration and forms the cornerstone of a country's mineral resource management. The mineral rights cadastre fulfills the following functions:

- Processing applications for various types of mineral licenses (prospecting, exploration, and mine development)
- Registering changes and updates to mining titles any time a title is granted, or ownership changes
- Checking license applications for possible overlaps with earlier claims or other impediments
- Advising the granting authority on whether a license application is technically admissible or not
- Ensuring compliance with payment fees and other requirements to keep a mining title valid
- Advising the granting authority when a mining title should be cancelled

The following core principles have been integrated into the legislation governing mining operations in most relevant countries:

- Mineral resources belong to the state.
- The right to explore and exploit the mineral resources can be temporarily transferred to an individual or a corporate entity through a written document, normally called a license or a lease.
- The mineral rights granted through such a license are considered real estate properties but are independent of surface or landownership rights.
- The holders of the granted license or lease must fulfill preestablished conditions to maintain their rights over the area.
- When the validity of the granted license or lease ends, the rights return to the state.

There are also several ground rules, or mining cadastre principles, that should be observed for a mining cadastre to operate properly:

- *Security of tenure.* This refers to the security of title, the right to transfer the title to any eligible third party, and the right to mortgage the title to raise money. It also refers to the transformation of an exploration license into a mining license. In most countries, mineral rights are divided into exploration and mining (or exploitation) licenses. Exploration licenses give holders the right to explore and evaluate the economic viability of mineral resources within the granted area. If an economic resource is confirmed, the exploration license must be transformed into a mining license for the holder to exploit it, provided the license holder has received all other necessary approvals and demonstrated compliance with the existing license terms and conditions.

  - *Security of title.* Licenses and mineral rights cannot be suspended or revoked except on specific grounds, which must be objective and not discretionary, and which must be clearly specified and detailed in the legal framework.

  - *"First come, first served."* Exploration licenses are granted on a first-come, first-served basis, which means that the first individual or company to apply for the rights to a certain area where mineral resources may exist will have the priority right to be granted the license or lease.

  - *Auctions.* Many countries include provisions in their mining law to auction available areas when resources or reserves have been assessed or inferred within an acceptable range of probability.

Efficient administration of the cadastral function requires the establishment of electronic databases. These list exploration and mining development sites, including all associated technical information, linked to geographic coordinates. All mining titles, their history, and specific attributes should be compiled into a modern mining cadastre, freely available to the general public via the Internet.

Overview of Extractive Industries Tax and Royalty Regimes

This section draws on material from Cameron and Stanley (2012), EI Source Book, chapters 6 and 7, as well as on other sources. For an in-depth treatment of EI taxation, readers are referred to Daniel, Keen, and McPherson (2010), The Taxation of Petroleum and Minerals. Readers interested in subnational sharing of revenue are referred to Anderson (2012), Oil and Gas in Federal Systems; and McLure (2003), Ahmad and Mottu (2003), and Brosio (2003).

EI fiscal regimes are composed of fiscal instruments common to all sectors, and those specific to the EI sector. The first category includes profit taxes, employment taxes, customs duties, value added tax (VAT), and dividend or interest withholding taxes. Fiscal instruments specific to the EI sector include mining royalties, oil production-sharing agreements (PSAs), sector-specific corporate income taxes, "windfall" or resource rent taxes, and EI-specific cost-recovery provisions. The goal of these fiscal instruments is to enable governments to capture a significant share of the rents generated from extraction, which is also referred to as the "government take" from extraction. This section provides only a brief overview of the relevant fiscal instruments.

Fiscal instruments in the EI sector can be assessed and compared on the basis of (i) the degree to which they distort economic activity, (ii) the ease of their administration/technical capacity requirements, and (iii) the extent to which they delay the onset of revenue flows. Profit taxes, royalties, resource rent taxes, and production sharing can be compared using these three parameters.

Profit taxes, such as corporate income tax, are often implemented at a separate and higher rate for the EI sector. They are nondistorting in the short run, in the sense that a project profitable before tax will also be so after tax. (All taxes are distortionary in the long run since they will affect returns to capital, which in the EI sector can have implications for the optimal economies of scale and the cutoff grade at which extraction will take place.) In other words, profit taxes are neutral (in the short run); tax revenues are proportionate to a company's profits. In terms of capacity requirements, profit taxes frequently require a level of administrative capacity that goes beyond what many developing countries' tax administrations can accomplish without external assistance. The administration of profit taxes requires skilled and qualified auditors able to assess company financial statements and capable of addressing issues such as abusive transfer pricing, thin capitalization, or the integrity of cost deductions. Because of the high initial capital investments prevalent in the EI sector—frequently combined with cost-recovery provisions to encourage exploration and development by investors—exclusive use of profit taxes may defer fiscal revenues for years, sometimes a decade or more. In some investor home countries, such as the United States and the United Kingdom, profit taxes paid in the host country qualify for foreign tax credit (whereas royalties do not), and are therefore considered attractive by companies from these countries.

Royalties may be levied either as per-unit tax (which is a uniform, fixed charge levied on a specified unit of production) or *ad valorem* (which is a fixed charge levied on the value of output, that is, gross revenues). Royalties are distorting (in the short and long run), in the sense that they may affect production decisions that are profitable on a pretax basis. The regressive nature of royalties (that is, the effective tax rate decreases as operational profitability increases) has prompted the renegotiation of contracts during resource booms, as governments have felt entitled to a share of the windfall profits. The same regressiveness can also prompt otherwise profitable projects to be prematurely abandoned during a commodity price slump, as operating income becomes negative when royalties increase relative to profits. In mature, high-cost oil basins such as those in Norway and the United Kingdom, royalties have

Figure 5.2 Stylized Representation of Volume-, Value-, and Profit-Based Taxes

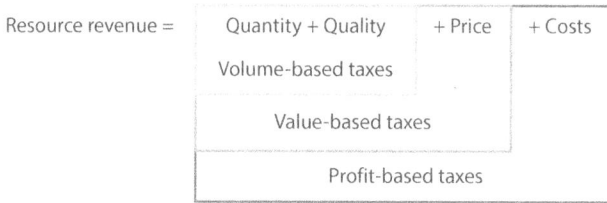

Resource revenue =	Quantity + Quality	+ Price	+ Costs
	Volume-based taxes		
	Value-based taxes		
	Profit-based taxes		

Source: World Bank.

progressively been eliminated (Nakhle 2010). Royalties are relatively simple to administer from an accounting perspective but require significant and highly specialized capacity for physical audit, including specialized skills in mineralogy to determine ore grades, volumes, and value (see figure 5.2). A major advantage of royalties, as seen by many governments, is that they provide revenue from the start of production.

Progressive tax instruments, or *resource rent* ("windfall," "rent," "additional profit") *taxes*, essentially add progressiveness to fiscal regimes dominated by (regressive) royalties. In a limited number of cases, countries have also tried to add progressiveness to otherwise neutral corporate income taxation, by linking the tax rate to profitability. Resource rent taxation targets returns on investment that exceed the minimum (risk-adjusted) reward necessary for capital to be deployed (Land 2010), or some determined threshold rate of return. Resource rent taxes have sometimes been imposed as a result of public pressure during commodity price booms to increase government revenues initially based on regressive royalties. Other instruments have also been used to capture windfall rents, such as sliding-scale production shares and sliding-scale royalties adjusted to price. Progressive tax instruments are in principle neutral, since they are designed to capture a share of perceived excess profits. Neutrality, however, requires a good proxy for profitability (to trigger the additional payment), which is not always achieved. For example, using production as a proxy ignores the price and cost, whereas using price as a proxy ignores production and cost. The most accurate mechanism would involve considering actual achieved company profitability. That, however, implies returning to the perceived administrative complexity of profit taxation, which may be an important reason why royalties are often preferred over profit taxes in the first place.

In the oil sector, *production sharing* is a frequent form of state participation (figure 5.3). Production sharing provides the state with an equity share income through ownership of production after cost-recovery by the private oil company, without offsetting financial obligations. In terms of its properties, production sharing is similar to profit taxation in that it is neutral and may require full cost-recovery before revenues start accruing to the state.

Bonuses are one-off payments linked to particular events such as a license award or signature, or to the attainment of a particular level of production.

Figure 5.3 Production-Sharing Agreements

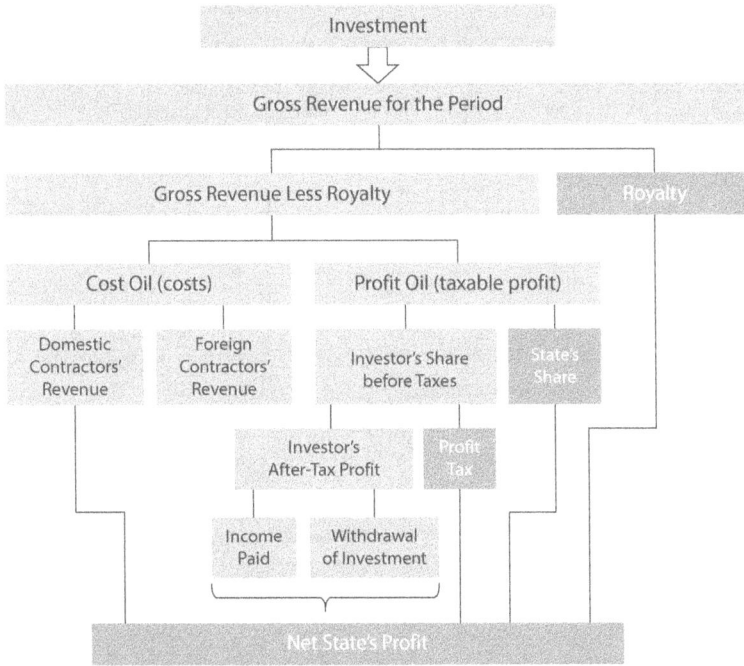

Source: Adapted from the Central Bank of Russia (2011).

Signature bonuses provide an early source of revenue, require very little administration, and may be perceived as particularly welcome in contexts where other significant revenues are years away (for example, where there is a long mine construction phase). Signature bonuses can run into hundreds of millions of dollars. They are neutral in the sense that, once paid, they do not affect investment or production decisions. Reservations about signature bonuses relate primarily to risk, both technical and political. On the company side, there may be concerns about the government's commitment to honor fiscal terms after the bonus is paid out. On the government side, the issue is the cost of the early revenue (represented by the bonus) in terms of foregone tax and royalty revenue in the medium and long term. In that respect, the signature bonus needs to be considered in the context of the overall EI fiscal regime and the net present value (NPV) of expected future revenue flows. In some cases, companies would prefer to pay out bonuses in several tranches, triggered by milestones that derisk the investment from both technical and permitting perspectives.

Other fiscal instruments relevant to the EI sector are capital gains taxes, withholding taxes, import duties, VATs, and tax holidays:

- *Capital gains taxes* may apply when licenses, concessions, or any asset changes hands, for instance, by being sold from a "junior" company with an appetite for risk to a more senior company in the event of a major discovery. It can be argued that to encourage exploration such sales should not be subject to taxation, and that the company that has made the discovery should recoup the full reward. But the large size of these types of premiums in recent years has prompted arguments that a share of these rents should go to the state.

- *Withholding taxes* on dividends and interest to shareholders or lenders allow the government to tax these types of flows, which would otherwise be difficult to tax once transferred abroad.

- *Import duties* are often waived for capital inputs in the project development phase, since these duties would essentially reduce tax revenues by imposing increased (deductable) costs on companies. Capital inputs at the production stage may nevertheless not be duty exempt.

- *Value added taxes (VATs)* are levied as a percentage of the value of goods and services, with VAT paid on inputs credited against VAT paid on domestic outputs. Since the produce from the EI sector is mostly exported, resource companies have little or no domestic output VAT against which to credit VAT on inputs.[1] Some governments have solved this issue by zero rating the VAT from domestic purchases destined for EI projects, as is the general practice for export sectors. When considering the VAT regime for the EI sector, it is necessary to avoid generating a bias against domestic producers whereby imports are duty free and local inputs are taxed.

- *Tax holidays*, once common, have now largely been discontinued in favor of less-distorting incentives, such as accelerated depreciation rates. Tax holidays were found to significantly reduce tax revenues by providing incentives for investors to accelerate the extraction of high-margin ores as much as possible before the tax holiday ended (see high grading, chapter 2, page 6).

Clear and appropriate definition of cost-recovery provisions is critical to the efficiency of the fiscal regime. Cost-recovery provisions, particularly in the oil and gas sector or wherever PSAs are utilized, are often notoriously underrated in terms of the attention they receive in the government's fiscal design processes (Cameron and Stanley 2012, section 7.2). Investors, on the other hand, are well aware of their importance. Cost-recovery provisions include a definition of recoverable costs, depreciation rates, cost oil limits for PSAs, investment and exploration expenditure uplifts, capital allowances, and limits on loss carry forwards. Some issues of frequent debate are the recovery of (i) overseas headquarters' costs (usually limited to a percentage of project costs), (ii) interest costs (subject to limits on debt-to-equity ratios, after which interest paid is subject to withholding taxes, and the application of market benchmarks if loans are to related parties), (iii) costs related to purchases from affiliated parties (addressed by applying the Organisation for Economic Co-operation and Development [OECD] rules on transfer pricing or requiring demonstration of third-party pricing equivalence [Cameron and Stanley 2012, 7.2]), and (iv) ring-fencing.

Ring-fencing establishes a limitation on cost-recovery across different activities or projects undertaken by the same taxpayer, so that costs associated with a particular license, mine, or oilfield are deducted from revenues generated within that field alone. Some countries ring-fence their mining, oil, and gas activities within a corporate entity, while others ring-fence individual projects. Without ring-fencing, resource tax receipts may be postponed if a company that undertakes a series of projects is able to deduct exploration and development costs arising from new projects against the income of projects that are generating taxable income. The absence of ring-fencing may discriminate against new entrants that have no income against which to deduct exploration or development expenditures (Nakhle 2010). As Mullins (2010) points out, ring-fencing also implies disadvantages. In certain jurisdictions, it may reduce the incentives for companies to undertake further exploration, if exploration expenditures cannot be deducted against income from existing projects that are generating profits. Ring-fencing may also encourage tax planning by locating less-taxed downstream production outside the ring-fence (including in another jurisdiction), or abusive transfer pricing that shifts costs outside the ring-fence to inside. Finally, ring-fencing can be complex if some taxes are ring-fenced while others are not. The choice of ring-fencing as a fiscal instrument thus requires governments to establish the necessary administrative capacity to address such issues.

Enhancing Competitiveness and Productivity

Oil, gas, and mining companies base their investment decisions on perceived risks, including those risks related to the business environment, political risk, and the size and magnitude of reserves. The higher the risk, the higher the potential future profits needed to justify the cost of investing in a new project. Equivalently, the higher the risk, the less favorable will be the conditions the company is willing to offer the government. Similarly, high costs of doing business, including overly complex or cumbersome legal regimes, will reduce the company's incentives to accept high tax burdens. Conversely, given comparable geological potential, investment capital will flow to countries with political stability, efficient and predictable policies and regulations, competitive tax rates, efficient infrastructure, and a skilled labor force. For example, a main reason that multinational oil companies operating in Norway have been willing to accept a sustained aggregate corporate profit tax for the oil sector of 78 percent is likely because political risk is considered to be extremely low, the business environment is favorable as well as highly predictable, and highly specialized labor is available locally.

In addition to the business environment parameters specified by the World Bank's "Doing Business" indicators, there are elements that are specific to the mining sector. Some recognized signatures of global good practice for mining legislation outline further parameters essential to the mining business environment (Gammon 2007). These include the following:

- *Clarity.* A mining law reflecting international best practices tends to be concise, with most of the detail being found within the regulations. Clarity serves to enhance investor confidence that there will be no bad surprises in the administration of the law, and reduces risk.
- *Transparency.* A transparent set of laws and regulations can enhance investor confidence. Minimizing the degree of discretion granted to state officials serves to improve transparency and reduces time-consuming and costly negotiations.
- *Completeness.* All relevant provisions should be gathered in one law, rather than being spread across several statutes. The full mining cycle should be covered, including geological surveys, exploration, development, feasibility studies, production, marketing, and mine closure and reclamation. This serves to avoid duplication, conflicts, and gaps between different laws.
- *Administration.* A technically competent ministry administering the laws inspires investor confidence. A well-funded geological survey can act as a tremendous stimulus to investor interest in geological potential. The state-of-the-art administration of mining titles (with details freely available on the Internet) is also greatly valued by potential investors.
- *Security of tenure.* A mining enterprise commonly involves many up-front risks and incurs substantial costs. Providing that an enterprise complies with all license requirements, it must have the right to move through the phases of exploration, development, and production.
- *Commodities.* For a mining company, it is the international norm to have the right to explore for, and produce, all minerals in a particular category (such as precious and base metals) found on the property.
- *Nondiscrimination.* There must be a level playing field for all prospective investors. When tendering concessions, all prospective investors should have an equal chance to win the tender. In the case of applications for exploration rights, the first-come, first-served principle typically prevails. Any deviation from these international best practices can cause a jurisdiction to rapidly lose investor confidence and suffer a competitive disadvantage.
- *Transferability of rights.* Various financing arrangements for the huge amounts of capital that can be required to bring a property into production often entail the entry of a new equity investor or lender, and having an ownership interest in the mineral rights can be an important consideration for such an investor.
- *"Use it or lose it."* For the state to ensure there are no undue delays in exploration, a system of exploration licensing should be developed that deters accumulation of large tracts of land, and forces the investor to relinquish land that is not needed. This frees up areas for other investors to acquire and explore.
- *Right of access.* In most countries, surface rights to property are privately owned, but subsurface mineral rights are state owned. To ensure the orderly and responsible exploitation of mineral assets to the benefit of all citizens, the

mining law should provide for exploration on private land. The need for strict and clear regulations that protect the rights of all parties is obvious.

- *Mineral data in the public domain.* To facilitate investor interest, detailed property-specific exploration data should be available in the public domain.
- *Internationally competitive fiscal regime.* The nature and rate of an EI fiscal regime can have a significant effect on the economics of an oil or mining project. The terms of the fiscal regime should be internationally competitive while providing an appropriate take for the state, and preferably on a sliding scale linked to commodity price cycles.
- *Interaction of ministry of mines with other ministries.* Other government ministries may have regulatory responsibilities that can significantly impact the mining sector. To maintain investor confidence in the overall regulatory regime, it is important that there be mechanisms and procedures in place to avoid conflicts between different regulatory provisions. Formal adoption of a national mineral policy can help ensure that laws do not conflict, but rather complement one another, in their application to the mining sector.
- *Approval deadlines.* It is important for the stakeholders that the mining project proceed in a timely fashion. The mining law and regulations should impose reasonable deadlines—on both the regulator and the investor—for application, approval, performance, and reporting activities.
- *Dispute resolution.* Clearly understood dispute resolution mechanisms and appeal procedures are essential for creating a favorable regulatory environment.

Note

1. In the event that the company sells its concentrate or product at a domestic mine or smelter it would collect output value added tax (VAT) and be able to offset this against input VAT.

CHAPTER 6

Monitoring and Enforcing Contracts: Legal Obligations and Institutional Responsibilities

Legal and Contractual Regimes

The legal obligations placed on a company operating in the extractive sector may be enshrined in a mining law, in an individually negotiated contract, or within model contracts with specified exceptions. In a pure permit regime, the government details all the major obligations for mining companies in legislation, and companies apply for permits for exploration and for extraction under the terms defined by law. This framework allows little scope for discretion in granting differential terms for companies involved in the extraction of the same mineral. In a pure contractual regime, the obligations of individual companies are negotiated individually and detailed in a contract often referred to as a mineral development agreement. In practice, most countries do not operate a pure permit or a pure contractual regime. Developing countries tend to have regimes that are on the contractual end of the spectrum.

The range of obligations that a typical mining or petroleum company faces under a stylized resource contract includes fiscal regimes, operational commitments, environmental obligations, worker safety obligations, and social obligations. The most relevant of these for economists and public finance professionals—fiscal, environmental, and social obligations—are described in this chapter. Table 6.1 summarizes the key contractual obligations, the key government entities responsible for monitoring and enforcement, other ministries and agencies whose cooperation may be required for effective enforcement, and, finally, some of the budgetary consequences associated with the inadequate monitoring and enforcement of these contractual obligations. For a checklist of guidelines on effective resource contract enforcement, see appendix D.

Table 6.1 Key Contractual Obligations: Enforcement and Budgetary Impacts

Type of contractual obligation	Nature of contractual obligation	Government entities with primary responsibility for M&E	Other government entities involved in M&E	Potential impact of ineffective M&E on the government budget
Fiscal regimes	Payment of taxes, royalties, license fees, and other forms of resource revenue	• Ministry of finance • Revenue collecting authority	• Ministry of oil and gas, or mining, responsible for verifying quantity and quality of resources, resource prices, and production costs • Customs and ports authorities for verification of quantity of resources exported	• Revenue collection losses that can impact budgetary execution • Overoptimistic resource revenue forecasts that feed into macrofiscal projections
Operational obligations	Submission and adherence to operational work plans to ensure that exploration and extraction occur in a timely manner	Ministry responsible for oil and gas, or mining	• Ministry of environmental management, if operational plans have environmental impacts • Ministry of employment and labor, if operational plans have employment-related impacts	Unrealistic medium-term forecasts for resource revenues on account of a lack of understanding of production risks posed by hurdles in the implementation of operational workplans
Social and community obligations	Community consultation; building and maintaining local infrastructure	• Ministry of mining • Relevant subnational government entities	• Central government entity responsible for subnational governments • Ministry of planning	• Inacurate assessment of expenditure liabilities arising from nonrecognition of recurrent maintenance costs for local infrastructure • Inaccurate assessment of contingent liabilities that may arise from lack of delivery of infrastructure at the local level
Environmental obligations	Submission of environmental impact assessments (EIAs) and submission and adherence to environmental management plans (EMPs)	• Ministry in charge of environmental management • Ministry of oil and gas, or mining	• Ministry of planning • Relevant subnational governments	Insufficient understanding of contingent liabilities arising from environmental damage, including potential costs for re-settlement and rehabilitation of displaced communities
Occupational and health safety obligations	Maintaining worker health and safety standards and ensuring workplace safety regulations	Ministry of oil and gas, or mining	Ministry of employment and labor	Limited direct budgetary impact, but noncompliance with these obligations may cause labor unrest and impact on production

Source: World Bank based on Smith and Rosenblum (2011). Courtesy of the Natural Resource Governance Institute.

Note: M&E = monitoring and evaluation.

Building Transparency and Accountability in Contract and Revenue Management

Timely and regular release of comprehensive information on resource licenses, contracts, and revenues allows governments, legislatures, and citizens to exercise oversight and conduct an informed debate about the best use of revenues (IMF 2007). Within the government, transparent procedures and accountability for their implementation, as well as transparency and integrity of the contract award process, reduce the level of discretion of individual government officials, enabling better oversight. This increases the likelihood that contracts will be negotiated and monitored according to government priorities. The public oversight enabled by revenue and contract transparency reduces the risk that revenues may be captured by narrow interest groups within the government without public scrutiny and debate.

The legal framework is the main pillar of resource transparency and provides the basis for reconciling the possibly divergent interests of key stakeholder groups. Stakeholders include the government, private investors, owners of land surface rights, local communities, and others that may be affected by the social and environmental impacts of the extractive industries (EI) sector. More specifically, revenue management is to an important extent dependent on relationships between the government; national resource companies (NRCs); and multinational oil, gas, and mining companies, since many transactions arising from these relationships have fiscal implications. For revenue flows to be transparent, roles and responsibilities must be clearly established in the legal framework (IMF 2007).

The clarity and openness of licensing procedures is a second pillar of transparency in the extractive sector and is essential to achieving revenue transparency during subsequent stages of mine or oilfield development. In the contracting phase, transparency refers to the availability of data describing contracts and licensing terms and procedures, as well as the regulatory regime that determines access to this information. The government's policy framework and legal basis for taxation or production-sharing agreements (PSAs) need to be made publicly available in a clear and comprehensive manner. Quasi-fiscal expenditures of NRCs, sometimes for social purposes or subsidies, should be clearly defined and described in the budget documents (IMF 2007).

The Extractive Industries Transparency Initiative (EITI) offers the benefits of transparency to governments, companies, and civil society by consolidating payments as disclosed by companies with receipt of payments as reported by governments. Here, governments benefit from adopting an internationally recognized transparency standard, thus demonstrating a commitment to reform and to combating corruption. This improves the potential for tax collection and generates increased trust and stability. Companies benefit from a level playing field in which competing firms are required to disclose the same information. They also benefit from an improved and more stable investment climate that allows for better engagement with citizens and civil society. Citizens and civil society benefit from receiving reliable information about the sector, as well as a multistakeholder platform where they can better hold the government and companies accountable (EITI 2013). Ideally, voluntary EITI commitments will be translated into legal requirements for revenue and contract disclosure.

Many resource-rich developing countries are not able to provide reliable, basic aggregate data on resource revenues. In that respect, computerized records of assessed and collected resource taxes will provide comprehensive accounting data on a cash and accrual basis. To simplify further, resource revenues can be paid into a single nominated bank account and swept daily into a treasury account at the central bank (sometimes called a consolidated fund). Following Calder (2010), the account needs to be audited on a daily basis by the government's chief accounting officer, for reconciliation with central bank records. The tax authority should contribute to transparency by preparing comprehensive accounts of taxes assessed, collected, and paid into the account.

To ensure integrity, the administration of resource revenues should be organized to minimize opportunities for collusion, without making procedures overly complex. Calder (2010) recommends that audit staff should for that reason not be involved in routine revenue assessment and collection, and the audit should be overseen by managers not directly related to the audit. Furthermore, he suggests that work on audits should be done by teams rather than individuals, and staff should be reassigned every few years. Appeals and reviews should be carried out by staff not directly involved in the decisions being reviewed, and the right of appeal should be ensured by an independent body. Information technology (IT) systems need to identify the staff member who enters the data, and these data should in turn be cross-referenced with the source document. Resource revenues should be audited by a body independent of the executive government, and this body should also address risks associated with administrative systems and procedures.

Monitoring and Enforcing Fiscal Regimes for the Extractive Sector

Interested readers are referred to Calder (2014), Administering Fiscal Regimes for the Extractive Industries: A Handbook; *and to Guj and others (2013)*, How to Improve Mining Tax Administration and Collection Frameworks: A Sourcebook.

To reach its potential for revenue collection, even a well-designed fiscal regime needs to be fully complemented by an effective revenue administration. Given the often very large revenue potential of the EI sector, an effective resource tax system has the potential to repay the costs of running the entire tax administration many times over.

Extractive revenue administration differs from regular tax administration in some important ways. The complexity of its administration arises chiefly from the complexity of the fiscal instruments typical of the sector (as described in chapter 5). This complexity is in turn driven by the nature of the industry and its economics (as described in chapter 2), such as nonrenewability, super-normal profits, high uncertainty and risk, long periods of operation, and price volatility.

Profit-based and other progressive taxes, while being more efficient than royalties in capturing rents arising from commodity price increases, tend to place a larger burden on tax administrations. Some countries therefore choose to adjust tax policy to administrative capacity. The other option is to adopt more sophisticated and progressive fiscal regimes and to address the administrative challenges by establishing long-term policies for capacity building, combined with qualified external support in the interim.

The best option for a resource-rich country with limited existing capability in revenue administration is still a matter of debate (Guj and others 2013), but there is broad consensus that, at present, extractive fiscal regimes worldwide tend to be too complex to be effectively administered (Calder 2014). Meanwhile, some resource-rich industrialized countries have simplified oil, gas, and mining tax codes, a process that greatly facilitates tax administration. Norway, for example, imposes a uniform 78 percent corporate profit tax on all oil-producing companies. In developing countries, by contrast, tax administration is often complicated by multiple individual extraction contracts, each with different terms and sometimes containing complex combinations of royalties, profit taxes, and windfall taxes. In the mining sector, some countries have reduced transaction costs by developing model mining agreements. Others have used tax codes to define universal regimes.

Challenges particular to the administration of EI sector revenue arise in the area of audit (Guj and others 2013). On the surface, an EI sector audit should be simpler than that of other sectors, since extractive industries are characterized by payments of very large contributions by few companies. This contrasts with most other tax administrations, where the number of tax contributors is far larger, but revenues from each contributor may be only a tiny fraction of what an oil or mining company contributes. But audits in the extractive sector are relatively complex and require high capacity (Calder 2014), as they frequently involve analyzing complex operations and transactions among several companies that belong to the same group, in addition to controlling the taxation affairs of the main shareholders. Resource companies may be related to other companies that are located in foreign countries, and relevant information may need to be requested from foreign tax administrations. Auditors responsible for extractive taxation also need to be familiar with practices such as thin capitalization, international profit shifting, and transfer pricing.

Another feature that introduces a considerable challenge to EI revenue administration is the need to involve several ministries and agencies across the government (Guj and others 2013). While in most other sectors, tax administration is exclusively the responsibility of the ministry of finance, the strong technical component of EI taxation requires the involvement of the EI sector ministry and NRC, where one exists. EI taxation requires that the production quantity and quality, the appropriate price, and the production costs be correctly determined—and, for each of these variables, the input of the EI sector ministry or NRC is critical (box 6.1).

Box 6.1　Establishing the Extractive Industries Tax Base: Generating Production Data

To establish the EI tax base, relevant government agencies and ministries must have the capacity to verify production quantity, quality, and cost, as well as the resource price:

- *Production quantity and quality.* Technical and laboratory equipment and capacity (for example, for mineralogy) are necessary to measure or verify the quantity and quality (ore or crude grade) of production, which, in turn, determines price. Since the sector ministry is often heavily involved in day-to-day physical inspection and regulation of the sector, the responsibility for inspecting ore quantity and quality follows naturally, with sample analysis frequently outsourced to independent laboratories. Similarly, in the case of production-sharing agreements (frequently used in the oil sector), the competencies of national oil companies are used to measure the quantity and quality of government profit oil, and for marketing and selling it.
- *Determining the resource price for EI taxation.* Relying on the sales price of resources reported by companies for tax assessment may present risks—especially in cases of transfer pricing between connected parties, whereby resources can be "sold" at a lower price to a connected company in a jurisdiction where the relevant tax is lower. To mitigate this risk, resource prices may be assessed based on prices listed on international exchanges, or by specialized firms that offer pricing services. But the prices of some commodities are not listed on international exchanges; furthermore, resource prices may vary depending on the quality of the resource and transportation costs. In such cases, the assessor may need to rely on the sector ministry to provide market intelligence and monitoring to establish credible free-on-board (FOB) prices.
- *Determining the production costs relevant to taxation.* Assessing production costs requires considerable skill and the cooperation of the sector ministry. Factors to consider include the treatment of stocks, provisions, and reserves; cost recognition; ring-fencing of costs related to non-production-related activities (such that they are excluded from calculations for resource taxes); incorrect categorizing of costs for the purpose of "uplift" (that is, inflating actual costs by a fixed percentage for tax deduction purposes); issues arising from thin capitalization, finance leasing, and currency gains and losses; cost-control rules and mechanisms (as agreed per the legal contract, as well as specific limits on the extent of deductions); and finally, the treatment of cost offsets, such as insurance recoveries, and the extent to which loss may be carried forward.

Source: Smith and Rosenblum 2011; Calder 2010.

Therefore, in order to be effective, an EI revenue administration requires coordination between EI sector ministries and the ministry of finance (Guj and others 2013). Such coordination often proves difficult in practice. To ensure communication, procedures for the exchange of information must be established, possibly set out in legislation, and included in staff job descriptions. Other possible measures include colocation of the sections of various agencies that are engaged in resource taxation, and staff exchanges or temporary secondments (Calder 2010). As discussed in chapter 3, in the section on NRCs, best practice suggests that NRCs not be involved in revenue management. If an NRC is already involved, good communication between the ministry of finance and the NRC is critical.

But the need for coordination among multiple government agencies does not imply that the responsibility for EI revenue administration should be divided across government. Such an approach has considerable disadvantages (Calder 2014) that include increased complexity; more regulators for companies to deal with; duplicated work; lack of clarity regarding responsibilities; lack of accountability; and uncoordinated management, systems, and procedures.

Calder (2014) therefore recommends minimizing the number of agencies in charge of EI taxation, and concentrating administration in a specialized unit. This is likely to strengthen revenue collection, since EI revenues come from a small number of very large tax contributors. The specialized unit can be separate or, if there are equally large contributors outside the resource sector, it can be located within a large taxpayer unit (LTU). The aim should be to make the specialized unit a center for administrative excellence (Calder 2010).

In the experience of many developing countries, setting up a special structure to control large taxpayer compliance has generated both increased compliance and more effective tax administration overall. According to Baer (2002), such advances have been achieved through improved knowledge of large taxpayers and their operations, more accurate and timely filing and payment of returns, the earlier detection of noncompliance through oversight of filing and payment obligations, more effective audits performed by better-trained auditors (focused on the EI sector), and reduced stocks of arrears. But to avoid the risk of a centralized revenue management system being used by powerful individuals for private gain, centralization must be backed by transparency and control measures.

The distribution of tax administration may also be regional; some countries may assign a portion of the responsibilities for resource taxation to the recipient region. It should be mentioned that the most effective LTUs tend to include strong and centralized supervision of operations, whereas highly decentralized units tend to be the least effective (Baer 2002). This principle is likely to be even more pronounced in the case of EI taxation, given the requirement for specialized technical skills.

Practices differ in the administration of nonresource taxes paid by resource companies (such as withholding taxes and value added taxes [VATs]). The concentration of all taxation of resource companies in one office is convenient for companies. On the other hand, to the extent that the administration of nonresource-specific taxes places a demand on the capacity of the LTU, it may distract attention from the main task of resource taxation. Calder (2014) therefore recommends that taxes not specific to extractive industries should in most cases remain under the regular tax administration regime, to avoid situations where the resource tax unit ends up dealing with a multitude of low-revenue taxes paid by the same companies. These taxes may be more efficiently administered by the units that usually administer such taxes.

Regardless of the institutional structure employed, for resource taxation to be effective, the often very large imbalances in expertise between governments and resource companies must be reduced by appropriate staffing and staff training (Guj and others 2013). Here, quality is more important than quantity. Only a small number of specialists are required to staff an oil, gas, and/or mining taxation office. The United Kingdom, with a relatively complex oil sector consisting of many small oilfields, has been employing only 50 staff for this purpose (Calder 2010). But the staff must be well qualified, trained, and equipped with the physical and IT infrastructure that their work requires. Importantly, to attract and retain staff of this caliber, the EI tax office needs to offer competitive pay and employment conditions (Guj and others 2013). This is particularly critical in a context where the government and the resource companies will often be competing for the most-qualified staff. Given the very large revenues frequently at stake in the EI sector, a strong case can be made for providing sufficiently favorable employment conditions for staff to seek employment and remain with the resource taxation office. However, experience shows that without consistent political and management support, in addition to adequate financing and qualified staff, even LTUs that are successful initially may succumb to the same problems that existed before their implementation (Baer 2002).

As discussed in chapter 3 on NRCs, many are tasked with some part of tax administration, in particular that of government profit oil. But due to conflicts of interest, NRCs' involvement in tax administration has frequently lacked accountability and transparency. A well-staffed NRC is therefore not a substitute for a qualified and capable tax administration. Meanwhile, favorable working conditions at an NRC may draw staff away from a tax administration tasked with the supervision of that NRC's operations.

Environmental Safeguards: Financial Sureties for Decommissioning

This section draws heavily on Sassoon (2009),"Guidelines for the Implementation of Financial Surety for Mine Closure." *Other relevant sources include Government of New South Wales (2010 and 2012), Peck and Sinding (2009), ProservOffshore (2010), Provincial Government of Western Cape (2005), and World Bank (2010c).*

Mining—and, to a lesser extent, oil and gas—projects have a large environmental footprint that, if not well managed, can leave the government with very large liabilities for damage to groundwater and soil. Such liabilities may run from a few million dollars for a small mine to more than a hundred million dollars for a large one (Natural Resources Canada 2012; Sassoon 2009). The pollution of groundwater (where previously suitable for human consumption and irrigation) can in the worst case saddle the government with mine liabilities running into hundreds of millions of dollars. It may cause health problems for the local population, make the groundwater inadequate for irrigation, and reduce the agricultural productivity of a region. Soil pollution may render an area unsuitable for agricultural production as well as for human settlement.

The "polluter pays" principle, as expressed in relevant sector legislation and regulation, protects the government against such liabilities. Thus, the resource company holds the responsibility for elaborating environmental impact assessments (EIAs), environmental management plans (EMPs), and decommissioning plans—the main tools of environmental management. For these tools to be effective, environmental ministries (as well as relevant sector ministries) need to have sufficient capacity and funding for environmental inspection and monitoring. Monitoring environmental compliance in the natural resource sector requires specialized staff and advanced equipment, including laboratory capacity for water and soil analysis.

Closure and decommissioning, the last phase of the extractive cycle, is the phase that holds the greatest risk of large liabilities. Mine closure is defined as the orderly, safe, and environmentally sound conversion of an operating mine to a closed state (Natural Resources Canada 2012). Aggregate liabilities for inadequately decommissioned mine sites may run into billions of dollars. Box 6.2 provides an example, from South Africa, of the large state liabilities that may result from inadequate decommissioning.

The objective of mine-closure activities is to restore the area impacted by mining to its original, premining environmental condition and to also address the social impacts of the mine's closure. Activities related to mine closure involve engineering works to decommission and dismantle infrastructure, complete rehabilitation, grade landforms for effective drainage, as well as the implementation of postclosure monitoring frameworks. There will be administrative work relating to the transfer of assets, labor force demobilization, and agreements for relinquishment of rights. The closure of operations in the oil and natural gas sector, known as "decommissioning and abandonment," often has social and environmental impacts that are similar to mine site decommissioning, but smaller in magnitude (Department of Energy and Climate Change 2011).

Box 6.2 South Africa: Large State Liabilities Resulting from Inadequate Decommissioning

A 2009 report from the auditor general of South Africa states that the estimated cost of rehabilitating South Africa's abandoned mines is R 30 billion ($2 billion). This amount was included in the financial statements of the Department of Minerals and Energy as a contingent liability in the fiscal years 2007/08 and 2008/09, although the long-term treatment of acid mine drainage and the construction and operating fees of plants was not taken account of in the R 30 billion. According to the Council for Geoscience (CGS), the costs related to the construction of these plants are likely to amount to as much as R 5 billion ($330 million), plus ongoing operating costs of several hundred million rand per year. Of the 5,906 abandoned mines, 1,730 were classified by the CGS as high-risk mines that would require approximately R 28.5 billion of the R 30 billion to rehabilitate.

Source: Adapted from the Auditor General of South Africa (2009).

A financial assurance, or surety, is the main financial tool used to ensure that environmental liabilities are not passed on to the government upon mine closure or oilfield decommissioning. Financial sureties are issued by financial institutions such as bonding companies, banks, and insurance companies. They ensure that the government holds sufficient funds to cover any costs that it may incur to achieve compliance with environmental requirements, in case the license holder is unable to meet these or fails to rehabilitate the environment. This can happen if an exploration or mining company goes bankrupt or defaults on its environmental commitments, and the administering authority is required to rehabilitate the area disturbed by the mining activities.

The amount of surety is defined as the maximum total rehabilitation cost for complete rehabilitation of all the disturbed areas (Queensland Government, undated, guideline A). The issuer of the financial surety ("security provider") agrees to be held liable for the acts or failures of a third party ("principal"), which in this case is the operating mining, oil, or gas company. If the company completes the decommissioning of the site of extraction in fulfillment of predetermined criteria, the value of the financial surety is returned back to the company in full, with deductions only for additional work that has to be completed by external contractors. Although the surety is normally the responsibility of the sector ministry, in some countries the ministry of finance has an important role in setting up and administering it. Figure 6.1 details the steps of managing financial sureties upon site closure.

Several types of financial sureties exist, including letters of credit, bank guarantees, surety bonds, trust funds, and cash deposits. Table 6.2 evaluates the comparative advantages and disadvantages of each of these instruments (MonTec 2007). The table also includes company guarantees, which are no longer in frequent use because they have been perceived as insufficient to ensure company compliance with closure requirements.

Figure 6.1 Managing Financial Sureties upon Site Closure: Four Administrative Steps

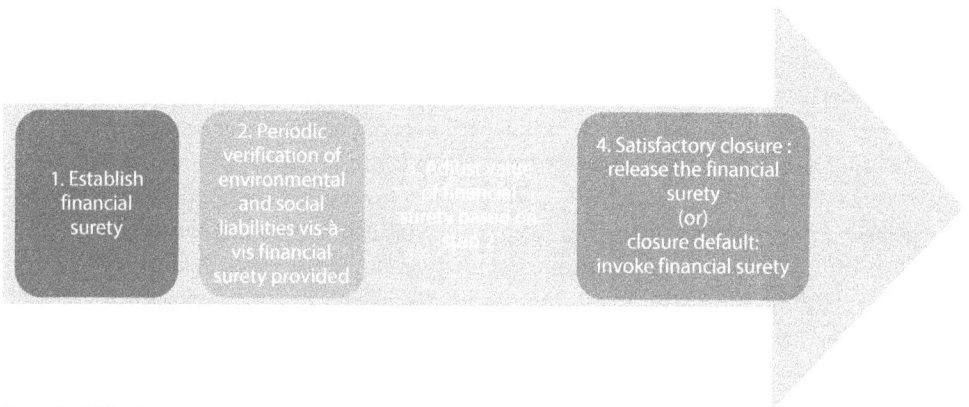

Source: World Bank.

Table 6.2 Evaluation of Commonly Used Financial Surety Instruments

Instrument	Advantages	Disadvantages
Letter of credit (LC), bank guarantee	• Cheap to set up (provided that the company meets the bank's requirements) • No tied-up capital • Modest cash outflow from mine operator • Few administrative requirements • The government can reserve the right to approve banks from which it will accept an LC, thereby minimizing the risk of failure of weak banks	• Surety provider (bank, surety company) may itself fail • Obtaining an LC may reduce the borrowing power of the mining company • Availability of bonds depend on the state of the surety industry and may be negatively affected by market forces outside the mining industry
Surety bond	• Generally low costs • No tied-up capital	• Bond issuer may fail over the long term • Ratings of the company determine the cost, which will be substantially higher for small companies (especially those without proven track records)
Trust fund	• High public acceptance ("visibility" of trust fund) • Trust funds may appreciate in value (but may also lose value, see "disadvantages")	• Risk of bad management of the trust fund (loss of value if fund invests in risky assets) • Trust fund may not have enough value accumulated through annual payments if mining project ceases prematurely • Trust fund management and administration consumes some of the value and income earned
Cash deposit	• Cash is readily available for closure and rehabilitation • Investment grade • High public acceptance ("visibility" of guarantee) • Can be used by small and junior mining companies, if they fail to meet the criteria of a bank	• Significant capital is tied up for the duration of the mine life, especially for large mining projects • Some governments may be tempted to use the deposited cash for purposes other than securing the mining project • Cash is vulnerable to being lost to fraud or theft
Company/ corporate guarantees	• Most advantageous for mining company • Does not tie up capital • Simple to administrate • Public availability of annual reports	• Annual reports and financial statements are not immune to manipulation (accounting scandals) • Self-guarantees have problematic public acceptance • Unless expressly mentioned as unconditional and irrevocable, enforcement of guarantees given by interested third-party companies might prove difficult, owing to possible challenges by the security provider, apart from the principal • Enforcement issues may arise if the security provider is a company that is registered abroad, or the assets secured are based abroad

Source: http://ec.europa.eu, copyright European Union, 1995–2015; MonTec 2007.

Social Safeguards: Community Foundations, Trusts, and Funds

This section draws heavily on several World Bank documents, including World Bank (2010b), Mining Foundations, Trusts, and Funds: A Sourcebook; World Bank (2010a), Mining Community Development Agreements—Practical Experiences and Field Studies; as well as World Bank (2011b), Sharing Mining Benefits in Developing Countries: The Experience with Foundations, Trusts and Funds. Interested readers are referred to the original documents.

Sharing the benefits of oil, gas, and mining projects at the local level is widely recognized as a necessity and has in many countries been made a legal requirement. The opening of a mine may lead to the displacement of the population in the concession area and propel societal change as migrant labor arrives from other regions. As a mine reaches the end of its productive life, the communities that have grown up around it may face unemployment and social disintegration. Like environmental issues, societal issues and their financial implications need to be taken into account from an early stage in the planning of an extractive project. Community development agreements (CDAs) in the oil, gas, and mining sectors contribute to the reduction of social tensions around the extraction process, allowing affected and otherwise eligible communities and regions to benefit from the extraction of natural resources from the lands they inhabit.

The control over local benefit sharing exercised by the central government varies from none to extensive. At one extreme, resource companies set up CDAs directly with local communities, with little or no government involvement. At the other extreme, the central government redistributes some share of the taxes and royalties collected from the resource company locally, and the investment of these revenues is implemented largely by the local government. Intermediate solutions place the central government in a regulatory role, in which it establishes and enforces procedures to ensure the appropriate selection, implementation, and evaluation of community investment projects financed in full or in part by resource revenues.

Although some companies still fund and implement CDAs by direct payment, most companies typically prefer to use some sort of community foundations, trusts, and funds (FTFs). This preference for FTFs reflects previous experience: direct payments have in some cases been channeled away from the intended beneficiary communities. Additionally, from a company perspective, FTFs offer the dual advantage of separating legal liability for community development projects from that of the company and delivering long-term, multiyear development projects that are cushioned from company budget cycles and commodity price fluctuations. Since FTFs are currently accepted as international best practice, this is the funding arrangement referred to in the rest of this section.

FTFs may be funded either directly by the resource company or through mineral tax revenues (transferred back to the mining region from the central government) or by a combination of the two. The collection of resource revenues for the implementation of CDAs depends significantly on the type and extent of subnational resource taxation or revenue sharing. In some countries, such as Papua New Guinea, Chile, and South Africa, resource companies are obliged by law to contribute a certain percentage of production or profits for community development, normally through FTFs. In Canada, company funding for FTFs is expected as a best practice, especially in areas belonging to indigenous communities, although it may not be formally required. In other countries, such as Peru, resource taxes and royalties are collected at the central level and then channeled via local government administrations to community development in the mining municipalities. Other countries collect resource taxes and royalties at the subnational level, some of which may be designated to fund local FTFs, or combine various types of revenue collection arrangements. Many countries grant tax exemptions to FTFs, since the goods and services provided by FTFs are essentially public goods.

The office of the auditor general has a natural role in monitoring and auditing if the government contributes directly to FTFs, if the government cofunds projects with FTFs, or if company contributions to FTFs are part of the company's obligations as specified by the mining contract. If FTFs receive only voluntary contributions and implement projects independent of local, regional, and national governments, it is in principle up to their contributors to ensure proper monitoring and auditing. An exception may be made where the local, regional, or national government makes nonmonetary contributions, such as land or permits that assume a specific social return.

Company contributions to FTFs can be set as a percentage of either production or profits. Production-based contributions tend to be preferred by communities and governments, precisely because they guarantee a financial contribution independent of company profits, whereas companies may prefer to have contributions set as a percentage of profits, or capital or operating expenditure, or based on an annual assessment of company funding availability. The advantage of using production and revenue as the bases for contributions is that they are relatively easy to verify. In some cases, smaller companies and exploration companies are required to contribute funds to a "pooled" community development fund, administered and controlled by the relevant provincial government. Box 6.3 provides examples of how the benefits of EI projects may be shared with local communities.

It is important to ensure that the funds administered by the FTFs are invested in such a way as to avoid undermining existing services or programs, and, wherever possible, aligned with and complementary to existing government and other initiatives. The roles of industry and government need to be clearly defined, to avoid outcomes in which the mining company takes the place of the government. Duplication of service delivery or investments, and hence of expenditure, can occur if project activities funded by the FTFs overlap with existing government service provision and investments. To best leverage resources, and to avoid wasteful duplication, FTF budgeting and expenditure management should ideally be linked with existing local and regional development plans.

Box 6.3   Financing for Community Benefit Sharing: Examples

There are several ways to share the benefits of extractive industry projects with local communities. These include through subnational revenue sharing, and community foundations, trusts, and funds (FTFs). Contributions to FTFs can be regulated or voluntary.

Subnational revenue sharing
Canon Minero Law, Peru

The Peruvian Canon Minero law requires that 50 percent of taxes paid by mining companies to the national government be channeled back to regional (25 percent) and municipal (75 percent) governments.

Tax and royalty sharing, Madagascar

Forty-two percent of taxes and royalties go to communes of extraction, 21 percent to the region, and 7 percent to the province.

National Equalization Fund, Senegal

Twenty percent goes into the fund. From this amount, local authorities from the mining regions receive 60 percent; the remaining 40 percent is shared by other local authorities in the country.

Community foundations, trusts, and funds (FTFs)
Newmont Ahafo Development Foundation, Ghana

Funded through a combination of 1 percent of net operational profits (before tax) from Newmont's Ahafo South mine, plus $1 per ounce of gold from Ahafo.

Raglan Agreement, Canada

Close to $16 million, or 4.5 percent of operating profits, went into a trust in 2007, which distributed the funding to three indigenous foundations that in turn distributed the funds to 14 communities in the Nunavik area of Northern Canada.

Source: Adapted from World Bank (2011b).

Government involvement can be crucial to success, but at the same time the government needs to be independent to act as a credible regulator. Some argue that regional and national governments should not be party to FTFs, since their regulatory and oversight function should remain objective, reserving government participation for the local level. It is frequently necessary for higher levels of government to be involved to some extent, however, especially if there is a lack of local managerial capacity. Governments can influence the structure and operation of an FTF, as when government officials serve on the FTF board, make compliance with local and regional development plans mandatory, or become involved in negotiations between communities and companies.

Building local investment capacity may be necessary to achieve desired results. Lack of sufficient organizational and technical capacity at the local level is a common, major challenge to the fruitful and productive implementation of FTF-financed projects. Experience shows that the most successful CDAs are those in which governments, resource companies, and/or nongovernmental organizations (NGOs) have invested considerable time and effort in building the capacity of all stakeholders prior to the start of the CDA. Capacity building addresses the skills not only of local government but also of community members, local organizations, and other stakeholder groups. The identification of potential partner organizations, with the necessary skills and experience to run broad capacity-building programs, will be a key component of successful capacity development. Typically, skills support at the local level will address capacity for strategic thinking and project prioritization, budget forecasting and analysis of costs, financial reporting and bookkeeping, project implementation and management, as well as monitoring and evaluation. Box 6.4 provides an example of developing local investment capacity, from Peru.

Box 6.4 Developing Local Investment Capacity in Peru

In the Cajamarca region of Peru, the Mineros Yanacocha mining company found itself trapped between the local population's expectations that its operations would make the Cajamarca region prosperous, and its inability to ensure that locals received substantial benefits from the public funds derived from the company's operations. The recipient municipalities of the community development agreements (CDAs) were not using the funds efficiently, and funds were starting to accumulate in the municipalities' accounts due to long delays in bringing projects to the implementation phase. Furthermore, the municipalities almost exclusively carried out smaller projects (such as paving roads) because they lacked the technical and managerial expertise to handle large, expensive projects. These problems were addressed at three levels:

- *Management of municipal investments.* Advisory services were provided, in cooperation with the International Finance Corporation (IFC), to help municipalities reengineer their organizational structure and build capacity to undertake small- and medium-scale infrastructure projects. External support was provided for large infrastructure projects, a model was provided for forecasting revenues, and options were presented to the municipalities for improving project execution.
- *Management of municipal finances.* Municipal staff were trained to improve their investment project budgeting and to leverage revenues from mining royalties to access other sources of financing, thereby increasing the funding available for local infrastructure projects.
- *Transparency and public feedback.* In cooperation with the University of Cajamarca and the local chamber of commerce, information on community development funds and municipal investments was disseminated, and surveys of these investments published.

---

box continues next page

Box 6.4 Developing Local Investment Capacity in Peru *(continued)*

Prior to this project, the mayors of the beneficiary municipalities felt that the CDAs were inadequate and did not recognize their own role as the authorities in charge of community funds. After the training was completed, the mayors recognized that they would be held responsible for the efficient use of the community development resources, and concentrated on making municipal and financial management more efficient. This program was later expanded into a national initiative called "Improving Municipal Investment," which included a Web site with course material and updates on municipal investment.

Source: Adapted from Aguilar and Francis (2005).

Ideally, the funds administered by the FTFs have, by the time the mining company ceases to operate, reached sufficient size to support both administrative expenses and project budgets throughout an extended period of decommissioning and closure. With this in mind, and given the time required to build the FTFs' endowment, company contributions to the endowment should start as early as possible to provide maximum benefits, although in the case of mining projects the high capital costs in the early stages mean that most mining FTFs are established after production has commenced. Under endowment arrangements, administrative costs are often sourced from the interest on the endowed investment, while periodic contributions from the mining operator (and/or other parties) support the development projects. Operating costs, including for transport, need to be considered in proportion to the amount of funding invested in development and are frequently capped at around 15–20 percent of total expenditure (World Bank 2011b).

CHAPTER 7

Public Infrastructure and Investment

From Subsoil Assets to Above-Ground Investment

Interested readers are referred to Rajaram and others (2014), The Power of Public Investment Management. *See also Ossowski and Halland (forthcoming, 2015), "Fiscal Management in Resource-Rich Countries," for a more detailed summary.*

The long-term growth of the extractive industries (EI) sector, and the replacement of natural wealth with public capital, depends on appropriate policies and solid procedures for public investment management (PIM). Investment decisions need to reflect economic and social costs and benefits, and assets need to be efficiently created, operated, and maintained. To avoid wasting natural resource revenues (figure 7.1), minimal PIM measures need to be established as quickly as possible (Rajaram 2012).

Improving public investment efficiency means improving project selection and implementation, as well as the operation and maintenance (O&M) of assets. This necessitates proper cost-benefit analysis carried out by well-trained government staff. The implementation stage requires solid procedures for budgeting and procurement to avoid inefficiency or, if relevant, corruption in procurement and delays in project completion (or incomplete projects). For maintenance purposes, public asset registries need to be established and updated.

The stages in PIM (figure 7.2) offer a framework for assessing the efficiency of public investment. The system can be scaled and adapted to different capacity contexts, but in all cases requires the establishment of authority and discipline in key institutions, as well as a political commitment to PIM. A key requirement is the establishment of a gatekeeping authority to reject ex ante projects that are poorly defined. Independent project reviews can be contracted to support the review of proposals and enhance the credibility of the investment process. A simple way to enhance investment efficiency is to track the rate of project completion. Public reporting increases incentives for ministries to improve their performance (Rajaram 2012). (The stages described in figure 7.2 correspond to the stages outlined in further detail in box 7.1.)

Figure 7.1 Revenue Leakages

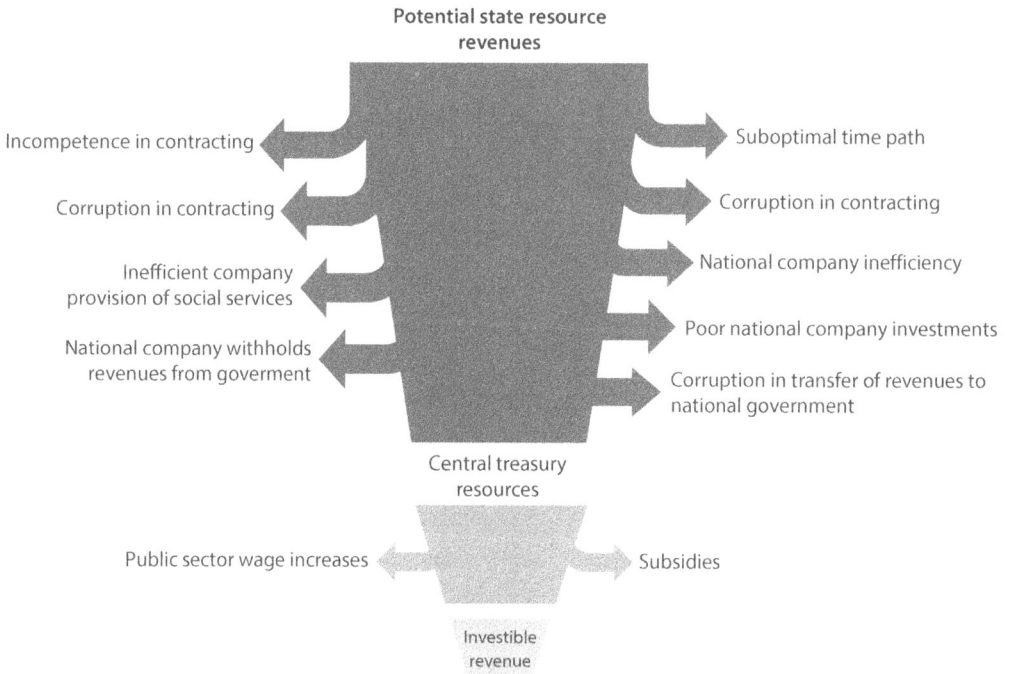

Potential state resource
revenues

Incompetence in contracting ←   → Suboptimal time path

Corruption in contracting ←   → Corruption in contracting

Inefficient company ←   → National company inefficiency
provision of social services

National company withholds ←   → Poor national company investments
revenues from goverment

                                                       → Corruption in transfer of revenues to
                                                          national government

Central treasury
resources

Public sector wage increases ←   → Subsidies

Investible
revenue

Source: Adapted from Ascher (2008) and Rajaram (2012).

Figure 7.2 Stages in Public Investment Management

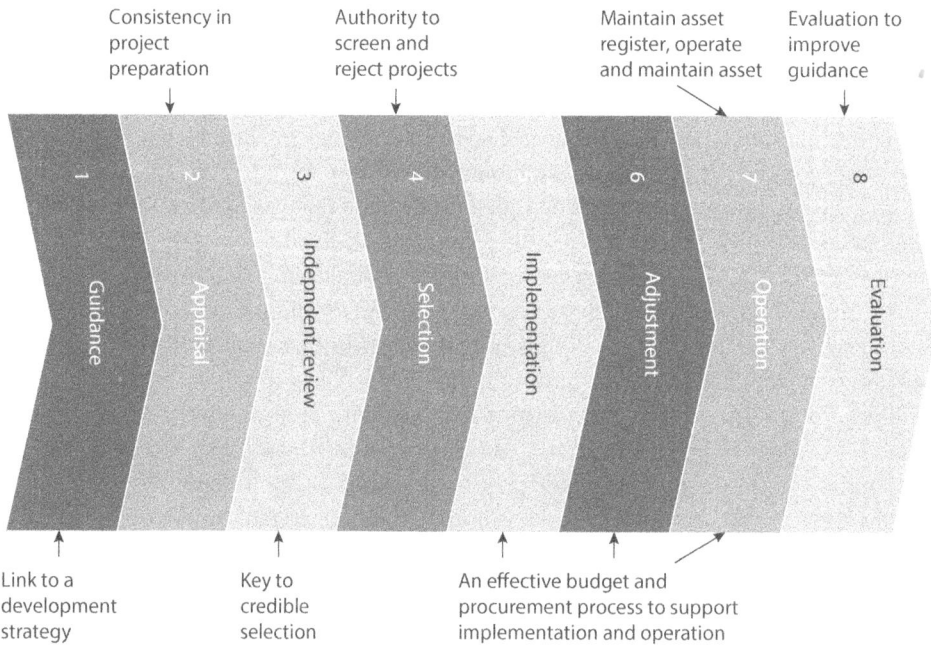

Consistency in   Authority to   Maintain asset   Evaluation to
project   screen and   register, operate   improve
preparation   reject projects   and maintain asset   guidance

| 1 | 2 | 3 | 4 | 5 | 6 | 7 | 8 |

Guidance   Appraisal   Independent review   Selection   Implementation   Adjustment   Operation   Evaluation

Link to a   Key to   An effective budget and
development   credible   procurement process to support
strategy   selection   implementation and operation

Source: Rajaram 2012.

Box 7.1   Effective Public Investment Management

High-quality public investment is essential to growth (Gupta and others 2011); poor invest-ment management may result in wasted resources and corruption. The risk of inadequate management is increased if investment is scaled up rapidly in the face of macroeconomic and institutional absorption constraints (Berg and others 2012). Efficient public investment man-agement (PIM) can be divided into four consecutive phases, each with several individual components (Dabla-Norris and others 2011; Rajaram and others 2010):

- *Strategic guidance and project appraisal.* Strategic guidance ensures that investment proj-ects are selected based on synergy and growth outlook and reflect development objectives and priorities. Projects that pass this first screening must then undergo scrutiny for financial and economic feasibility and sustainability. This requires several steps—financial and eco-nomic cost-benefit analyses, prefeasibility and feasibility studies, and environmental and social impact assessments—all undertaken by staff trained in project evaluation. Further-more, creating potential project lists strengthens accountability.
- *Project selection and budgeting.* Vetting proposed projects requires a politically indepen-dent gatekeeping function. The participation of international experts, together with national technical experts, can enhance the quality and robustness of the review. Link-ing the process of selecting and appraising projects to the budget cycle is necessary to account for recurrent costs, and to ensure appropriate oversight and consistency with long-term fiscal and debt management objectives. This requires a medium-term fiscal framework that translates investment objectives into a multiyear forecast for fund and budget aggregates.
- *Project implementation.* This covers a wide range of elements, including efficient procure-ment, timely budget execution, and sound internal budgetary monitoring and control. Clear organizational arrangements, sufficient managerial capacity, and regular reporting and monitoring are essential to avoid underexecution of budgets, rent seeking, and corrup-tion. Procurement must be competitive and transparent and include a complaints mecha-nism to provide checks and balances and a credible internal audit function.
- *Project audit and evaluation.* Ex post evaluation is in many developing countries a missing feature of PIM systems, as are adequate asset registers. Registers are necessary to maintain and account for physical property and should be subject to regular external audits.

Sources: Adapted from Berg and others (2012) and Dabla-Norris and others (2011).

Infrastructure Investment

An in-depth treatment of dual-use mining infrastructure is provided by the International Finance Corporation (IFC 2013) in "Fostering the Development of Greenfield Mining-Related Transport Infrastructure through Project Financing." For resource-financed infrastructure (RFI) deals, see Halland and others (2014), "Resource Financed Infrastructure: A Discussion on a New Form of Infrastructure Financing."

Extractive industries (EIs) have traditionally been an enclave sector, but the large infrastructure investments taking place here can be leveraged for public use and for wider private sector development. In some developing countries, the infrastructure built by resource companies constitutes a significant share of total infrastructure investment in the country overall. The mining sector, with its needs for high-capacity road and rail links, typically includes infrastructure investments that may run into the billions of dollars in the development phase and that frequently constitute the largest share of investment expenditures. IFC (2013) explores public–private partnership (PPP) models that allow for the dual use of this mining infrastructure, so that it contributes to the overall road, rail, port, power generation, and other infrastructure capacity of the country.

For investment in EI infrastructure to contribute optimally to a country's overall infrastructure capacity, it must be coordinated with the existing infrastructure and with plans for future regional and national infrastructure development. Some level of regional planning is in this respect essential, so that cross-border issues can also be taken into account. Territorial planning is needed to ensure that infrastructure grids are designed to optimally serve the public interest, while retaining the main, private purpose of serving the mine. In some cases, a government may coordinate the infrastructure investments undertaken by different companies, thereby providing not only a public good but also benefiting the companies involved, through economies of scale.

Leveraging extractive-related infrastructure for wider private sector development requires a comprehensive approach. A resource corridor or industry development zone, for example, can be used to address local industry demand for public goods such as power and transport. A resource corridor might make use of EI infrastructure to link local firms with demand from resource companies and to markets in general. In this regard, it is critical to establish programs for business regulation, small and medium enterprise (SME) financing, business development, and human resources (HR) development. Special regulatory regimes for resource corridors or industry development zones may be needed where it is not possible to efficiently address the business environment of the country as a whole. SME financing solutions may need to be developed for local companies that attempt to enter oil, gas, and mining supply chains; downstream production; or the sidestream provisioning of engineering, legal, and accounting/audit services, among other areas. Legal and institutional constraints on access to finance may need to be addressed.

In recent decades, resource-rich developing countries have been using their natural resources as collateral to access sources of finance for investment, countervailing the barriers they face when accessing conventional bank lending and capital markets. One of several financing models to emerge as a result is the resource-financed infrastructure (RFI) model, described in box 7.2, a derivation of previous oil-backed lending models pioneered by several Western banks in Africa.

Box 7.2   A Discussion of Resource-Financed Infrastructure

Resource-financed infrastructure (RFI) deals have been employed by governments to exchange resources directly for turnkey infrastructure. RFI establishes a direct relationship between (i) the government's future revenue stream from the resource component and (ii) a nonrecourse loan from the resource developer's lender or another financial institution, to the government for the purchase of infrastructure. The loan is paid down with the committed future government revenues from the oil or mineral extraction as part of the established fiscal regime. Loan disbursements for the infrastructure component are paid directly to the construction company to cover construction costs. The key to RFI is creating the nonrecourse link, using a special loan mechanism, between the committed future resource revenues and the current infrastructure financing.

According to Collier (in Halland and others 2014), RFI contracting may under circumstances of weak fiscal discipline represent a commitment mechanism that enables ministers to ensure that future decision makers devote a sensible proportion of resource revenues to the accumulation of assets. Unlike resource-backed lending or sovereign bond issuances, RFI connects government revenues from resource extraction directly to infrastructure investment. RFI thereby serves to countervail barriers to international capital markets and to bypass capacity constraints that governments face when required to implement large infrastructure projects. Another reason that existing versions of RFI transactions have been seen as attractive by governments may be that the RFI type of transaction is perceived as an opportunity to provide fast returns to citizens while decision makers are still in office. Since mines and oilfields take a long time to develop, the infrastructure could be in use long before the extractive project generates revenue or turns a profit.

Despite its potential benefits, RFI also brings significant risks and challenges. Early deals of the RFI type have generally been concluded on a noncompetitive basis, with little transparency or attention to structuring the transaction as a true financing model. This has brought up questions related to the valuation of the deals—how much infrastructure now for a certain amount of oil or minerals in the future? There have also been concerns with regard to the quality of the completed infrastructure, as well as the capacity for operation and maintenance—issues that in a mature RFI deal would be addressed through careful contracting, due diligence exercises, and independent third-party construction supervision. According to Wells (in Halland and others 2014) RFI deals should be evaluated like any other business arrangement, be carefully compared against alternative ways of obtaining returns from natural resources or financing infrastructure, and should include appropriate safeguards and procedures for implementation.

Source: Halland and others 2014.

CHAPTER 8

Economic Diversification and Local Content Development

Developing Linkages

Readers interested in local content are referred to Tordo, Warner, and Anouti (2013), "Local Content Policies in the Oil and Gas Sector."

As has been noted, the extractive industries (EI) have traditionally functioned as enclaves that—when operating in developing countries—often bring staff, goods, and services from abroad, with limited spillover to the domestic private sector. With important exceptions (Norway and Canada, among others), the extractive sector has in many countries not generated a high share of direct employment for locals and has created few links to local firms. Local firms in resource-rich developing countries frequently find it difficult to provide inputs into the EI sector production process, being hindered by cumbersome business regulations, lack of access to qualified staff, lack of management skills, lack of knowledge about international product standards, and lack of access to finance. The enclave nature of the EI sector has limited economic diversification away from the sector in many resource-rich developing countries.

To counter this trend and further economic diversification, policies may enhance EI sector linkages with the rest of the economy (figure 8.1). These linkages can take five main forms (Jourdan 2014):

- *Fiscal*, through the collection of fiscal revenues from the EI sector and subsequent public investment in physical and human capital
- *Spatial*, to promote the use and impact of EI infrastructure for public use and wider private sector development (as discussed in chapter 7)
- *Knowledge*, through human resource development and research
- *Backward*, through the participation of local firms in the EI production process
- *Forward*, through the promotion of the value addition of extractive commodities

Figure 8.1 Connecting Extractive Industries with the Larger Economy: Five Types of Linkages

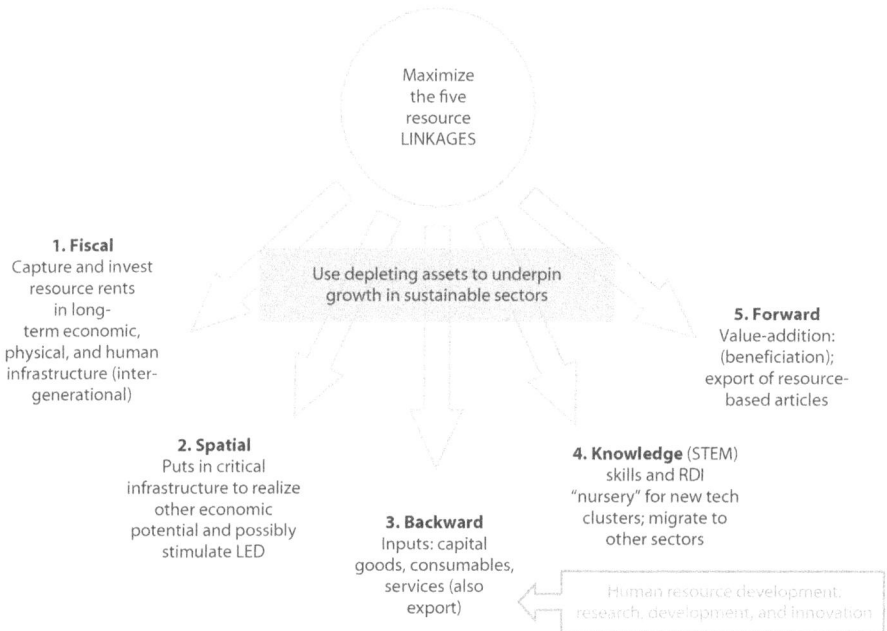

Maximize
the five
resource
LINKAGES

1. Fiscal
Capture and invest
resource rents
in long-
term economic,
physical, and human
infrastructure (inter-
generational)

Use depleting assets to underpin
growth in sustainable sectors

5. Forward
Value-addition:
(beneficiation);
export of resource-
based articles

2. Spatial
Puts in critical
infrastructure to realize
other economic
potential and possibly
stimulate LED

3. Backward
Inputs: capital
goods, consumables,
services (also
export)

4. Knowledge (STEM)
skills and RDI
"nursery" for new tech
clusters; migrate to
other sectors

Human resource development;
research, development, and innovation

Source: Adapted from Jourdan (2014).
Note: LED = local economic development; STEM = science, technology, engineering, and math; HRD = human
resource development; RDI = research, development, and innovation.

A number of resource-rich countries have sought to develop backward production linkages, or local content. Location advantages and potential technology spillovers from multinational oil, gas, and/or mining firms and their suppliers provide countries with a case for maximizing domestic upstream linkages and providing sidestream services. Companies can improve efficiency and decrease costs by using local producers. The use of local labor tends to be significantly less expensive than that of expatriate staff, and local supplies can be delivered sooner than those from afar. For these and other reasons, oil and mining companies may prefer to outsource their noncore activities (for which they do not have a comparative advantage) to local firms where possible. But there are significant obstacles to efficient local content development. For instance, firms in some areas may be unable to meet EI sector requirements, perhaps because they lack the human resource capacity needed to produce complex goods and services up to industry standards.

Knowledge linkages from the EI sector are critical to economic diversification. Many of the skills needed in the natural resource sector are transferable and can form the foundation for the development of other sectors. Examples include skills and training in mechanical engineering, civil engineering, electrical engineering, chemistry, accounting, finance, business management, and law, as well as technical trades such as welding, mechanics, and electrical installation. Raising the level of domestic knowledge and skills in the natural resource sector raises the general level of human capital (see box 8.1 for the case of Norway) and is likely to improve the capacity of resource-rich developing countries to absorb technology transfer in areas outside the natural resource sector.

Box 8.1 The Diversification of Norway's Oil and Gas Value Chain

In the early 1970s, Norway had very low levels of local content. The lion's share of goods and services for Norwegian offshore oil production were delivered by U.S.-based global oil and service companies. Three decades later, Norway had attained local content levels above 50 percent for capital inputs and above 80 percent for operation and maintenance (O&M) inputs. Based on the skills and competitiveness acquired in delivering local content, Norwegian firms supplying the oil sector achieved export rates of 46 percent of their sales.

To promote the development of local capabilities, international petroleum companies were, in the early phase of oil extraction, encouraged to enter cooperative agreements with research units at national universities. This resulted in the upgrading of oil-sector-specific skills among academic staff and degree programs tailored to the oil sector and related industries. The policy was characterized by a well-articulated system of evaluating operator contributions to domestic capacity. Financial support for research and development (R&D) was taken into account in the award of contracts, as was the transfer of skills and technology. A corporate income tax rate for the oil sector of 78 percent, with all R&D expenses immediately deductible, provided a strong incentive for investment in domestic R&D.

Similar policies were established at the firm level, encouraging multinational oil companies to integrate domestic firms and enterprises in large development projects and fostering joint ventures and cooperation agreements between domestic and foreign companies. International oil companies were required to set up fully operating subsidiaries in Norway. Such arrangements allowed domestic firms, many of which specialized in ship building, to convert their existing capabilities into capabilities for offshore oil production.

When selecting sectors for diversification, Norway chose sectors for which related capabilities already existed—that is, capabilities that could be used for activities such as building offshore oil platforms and delivering specialized technical services to the oil sector. Norway did not diversify significantly downstream, except for oil refineries and gas-processing plants, since the petrochemical industry was not perceived as a sector in which Norway would have a competitive advantage.

Sources: Adapted from Noreng (2005) and Heum (2008).

Countries have shown increasing interest in promoting forward linkages through so-called beneficiation policies that promote value addition in exports of EI commodities (despite limited evidence that such policies can be successful in isolation). In the mining sector, beneficiation policies have involved controls on exports of unprocessed commodities and also the promotion of downstream industries (such as steel manufacturing). A recent example of such a policy is the ban on unprocessed nickel and copper set by Indonesia in early 2014; this led to significant losses of export earnings and put upward pressure on nickel prices in contrast to other metals (Nair and Lee 2014). Empirical evidence on the efficacy of beneficiation policies suggests that they achieve limited success in promoting value addition in exports (Hausman, Klinger, and Lawrence 2007). Furthermore, the capabilities required for the development of downstream industries are distinct from those of the primary extractive industry, and thus, the presence of extraction activity in a country does not necessarily imply a comparative advantage for the development of downstream processing and manufacturing. This distinction is further highlighted in box 8.2, which provides an overview of the salient issues relevant to mineral beneficiation.

It is possible to enhance growth performance by focusing on mining and the export of unprocessed minerals—and the use of associated revenues to make productivity-enhancing investments. Chile's experience in managing the copper

---

Box 8.2 International Experience in Promoting Downstream Mineral Processing

Policies to encourage downstream mineral processing take various forms, from restricting exports of unprocessed commodities to providing subsidies for downstream processing and refining industries. For example, South Africa has imposed export controls on many unprocessed minerals and created financing programs to promote value addition in mineral industries. Several other African countries have taken a similar path, such as Botswana in diamonds, Zambia in copper, Ghana in oil, and Mozambique in natural gas and coal. In Australia, tax incentives and energy subsidies have been used to promote the downstream steel industry.

While the policy of promoting forward linkages in the mineral sector has gained popularity, there remains considerable debate on its impact, with a recent cross-country empirical study (Hausman, Klinger, and Lawrence 2007) failing to find positive effects of promoting value addition in exports. Across a wide range of industries, shrinking transport costs have driven a general trend toward global fragmentation of supply chains. This trend is also seen in the mineral sector, where only a very small number of countries that export unprocessed minerals also export the same minerals in processed form (figure B.8.2.1). Hausman, Klinger, and Lawrence (2007) investigate the efficacy of mineral downstream-processing policies on improving value addition in exports using trade data for the period 1975–2000 for all

box continues next page

Box 8.2 International Experience in Promoting Downstream Mineral Processing (continued)

Figure B.8.2.1 Global Copper Production, Refining, and Consumption Trends, 2013
million metric tons

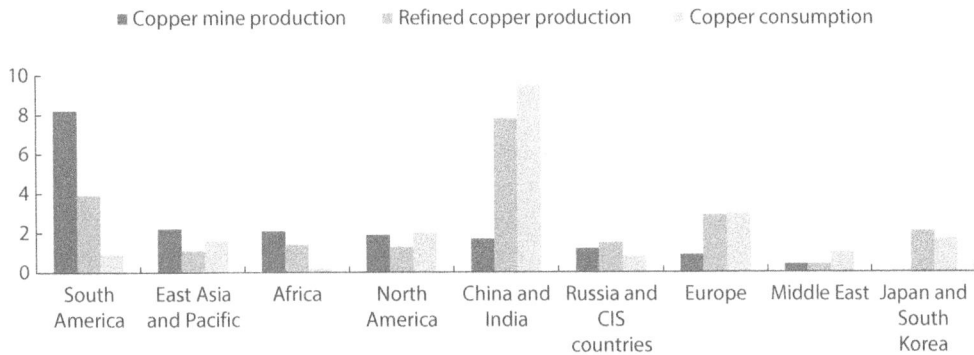

Sources: OECD 2010; SEADI 2013; Johnson and Noguera 2012; Hausman, Klinger, and Lawrence 2007; U.S. Geological Survey of Metals and Minerals 2013. Figures for copper from Indonesian Ministry of Energy and Mineral Resources.
Note: East Asia and Pacific excludes China, Japan, and South Korea. CIS = Commonwealth of Independent States.

countries and input-output data describing supply chain linkages for 241 products. They find that increases in value added in primary commodities are not associated with increases in the share of value added in the country's export basket in the medium to long term.

Mining and mineral processing are distinct industries requiring different capabilities; a country with a major mining sector may not be able to move profitably into downstream processing. Hausman, Klinger, and Lawrence highlight that the development of mineral-processing capabilities in a country has historically been linked to comparative advantages in other factors (such as energy), rather than access to raw mineral ore. Key determinants of the location of smelters or refineries for many minerals revolve around the need for complementary inputs such as low-cost power, access to land, pollution controls and other regulatory requirements, access to low-cost finance, and external economies (such as markets for by-products and so on).

industry is a case in point. Chile is the world's largest exporter of copper, accounting for over 40 percent of total global exports in 2012, but since the 1980s Chile has not focused on processing copper within the country and exports the bulk of its copper as concentrate to China and India, where it is smelted further. The experience of Chile is discussed further in box 8.3.

Countries have chosen various routes of extractive-led economic diversification, with variable levels of success. Local content development has been successful where countries have managed to develop the requisite conditions for national companies to evolve into internationally competitive commercial actors in the sector. EI sector policies aimed at promoting downstream value addition

Box 8.3   Institutional Infrastructure for Nonresource Diversification in Chile

Rather than become a major industrial manufacturer and exporter, Chile chose to develop a dynamic and more diversified commodity export sector based on its diverse resource base. An important element of Chile's success has been its efficient macroeconomic stabilization. Chile has also placed great emphasis on improving the business environment, and its "Doing Business" indicators are now ranked the highest in Latin America. Furthermore, since the 1980s Chile has implemented a number of "horizontal" or sector-neutral industrial policies, including access to finance for small and medium enterprises (SMEs). This issue is addressed by guaranteeing a certain percentage of the credit granted to SMEs by private sector financial institutions, by means of the Small Businesses Guarantee Fund. Another essential horizontal public good is provided by the export promotion agency, ProChile, which undertakes generic export promotion independent of sectors.

A 3 percent royalty from mining is earmarked for business innovation, with users being private firms, the semipublic Fundación Chile, as well as a consortia of universities and private firms. Innova Chile, a government entity that encourages business innovation, undertakes a range of activities including (i) technological assistance to SMEs, consisting of grants for the adoption of new technologies; (ii) grants to business associations, centers, and foundations for the development of norms and technological infrastructure; (iii) support for nodes, which are small administrative offices that seek to connect the technological needs of firms with domestic or foreign suppliers; (iv) support for business incubators; (v) support for gazelle firms (firms that have the potential for explosive growth); and (vi) business innovation, including both support for innovation by individual firms and grants to a consortia of firms and university institutes, for the development of new technologies.

Fundación Chile provides funding for projects that are riskier than those in already-established sectors and which the private sector is unlikely to fund on an optimal scale. The foundation is credited with adapting Norwegian salmon cultivation technology to Chilean conditions, a spectacular success that resulted in a major new export product. The process by which Fundación Chile discovers, develops, and converts an idea into a business opportunity is as follows: In the first stage, a product, technology, or service is identified that is profitable abroad and is not in use in the national economy. At the second stage, the production technology (and its license, if relevant) will be acquired and adapted to the domestic environment through research and development. If this indicates that the product or process is commercially feasible, production is scaled up in accordance with product characteristics and technology requirements.

Sources: Agosin, Larrain, and Grau 2010; Benavente 2006; Gelb 2011.

have in some contexts had limited success in ensuring the ultimate goal of enhancing EI sector linkages to the wider economy. As the case of Chile demonstrates (box 8.3), broader institutional infrastructure may be requred for successful nonresource diversification.

APPENDIX A

Resource Classification Frameworks

The Four Classification Codes

The terms "resources" and "reserves" are codified in (i) the securities exchange laws of most countries that trade in the securities of extractive industries (EI) corporations and (ii) systems of national accounting. Four main classification codes are today used to report volumes of natural resources (following a reform of reporting standards that started in the early 1990s). The classification code of the Committee for Mineral Reserves International Reporting Standards (CRIRSCO) focuses on minerals, while that of the Society of Petroleum Engineers–Petroleum Resources Management System (SPE-PRMS) classification codes focuses on hydrocarbons. Meanwhile, the United Nations Framework Classification for Fossil Energy and Mineral Reserves and Resources 2009 (UNFC-2009) and the System of Environmental-Economic Accounting 2012 (SEEA-2012) apply to all types of both mineral and hydrocarbon resources. None of these classification systems takes only geological criteria into account; economic and technical criteria are also considered. This implies that stocks of geological material have to be regularly reassessed in the light of new geological knowledge, progress in extraction technology, and shifts in economic, legal, and political conditions (OECD 2014). Each of the classification systems is discussed in turn below, and this discussion draws heavily on recent, comprehensive work of the Organisation for Economic Co-operation and Development (OECD 2014).

Committee for Mineral Reserves International Reporting Standards (CRIRSCO)

The CRIRSCO framework applies specifically to mineral resources and is consistent with the reporting requirements of major mining countries and of stock exchanges where mining companies are trading. The CRIRSCO framework captures the bulk of mineral resources and reserves that are reported by publicly listed mining companies. The total value of those mining companies listed on stock exchanges with CRIRSCO-compatible accounts is more than 80 percent of the listed capital of the mining industry (CRIRSCO 2015).[1]

As shown in figure A.1, the CRIRSCO classification system is two dimensional: the vertical axis is for "geological confidence," while the horizontal axis is for "modifying factors" corresponding to several socioeconomic factors, such as resource prices and legal constraints.

Care should be taken to never confuse resources and reserves—the reporting codes allow for both inclusive and exclusive reporting of reserves within resources.[2] The CRIRSCO code divides mineral resources into inferred, indicated, and measured categories based on the confidence with which the geologic characteristics of the resource are known. Further, measured and indicated resources can be *converted (or subclassified)* into proved or probable reserves without additional geological information, if the modifying factors (mining, processing, metallurgical, infrastructure, economic, marketing, legal, environmental, social, and government) indicate the deposit can be profitably mined.

Declaring a mineral reserve requires the qualified person to make reasonable projections about the future. A number of the modifying factors will vary over the period of planned depletion (from, say, 5–20 years or more). As a result, whether a particular mineral deposit is considered economically feasible depends on the time horizon and assumptions about future costs, prices, laws and regulations, and

Figure A.1 CRIRSCO Framework for Mineral Reserves and Resource Classification

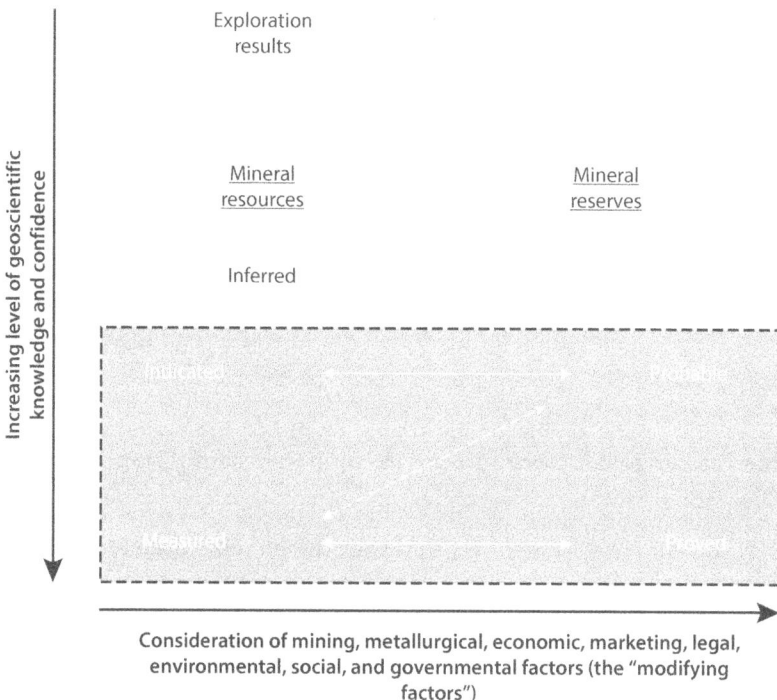

Source: Adapted from CRIRSCO (2015).

social and environmental policies. Some of the uncertain modifying factors could be evaluated using statistical or engineering cost models developed for regions with similar cost structures. Other modifying factors would need to be resolved on a country-specific basis for country policies related to taxes, royalties, social policies, and environmental regulations. The subcategories under the CRIRSCO framework are defined in box A.1.

The following definitions come directly from CRIRSCO (2013). Readers are encouraged to visit the CRIRSCO Web site, as definitions may change.

Box A.1   CRIRSCO Classification System Definitions

A *proved mineral reserve* is the economically mineable part of a measured mineral resource. A proved mineral reserve is accompanied by a high degree of confidence in the modifying factors. A proved mineral reserve represents the highest confidence category of a reserve estimate.

A *probable mineral reserve* is the economically mineable part of an indicated or, in some circumstances, a measured mineral resource. The confidence in the modifying factors that apply to a probable mineral reserve is lower than that applying to a proved mineral reserve.

A *measured mineral resource* is that part of a mineral resource for which quantity, grade or quality, density, shape, and physical characteristics are estimated with confidence sufficient to allow the application of modifying factors to support detailed mine planning and final evaluation of the economic viability of the deposit. Geological evidence is derived from detailed and reliable exploration, sampling, and testing and is sufficient to confirm geological and grade or quality continuity between points of observation. A measured mineral resource is accompanied by a higher level of confidence than is either an indicated or an inferred mineral resource. It may be converted to a proved or to a probable mineral reserve.

An *indicated mineral resource* is that part of a mineral resource for which quantity, grade or quality, density, shape, and physical characteristics are estimated with sufficient confidence to allow the application of modifying factors in sufficient detail to support mine planning and evaluation of the economic viability of the deposit. Geological evidence is derived from adequately detailed and reliable exploration, sampling, and testing and is sufficient to assume geological and grade or quality continuity between points of observation. An indicated mineral resource is accompanied by a lower level of confidence than is a measured mineral resource and may be converted to only a probable mineral reserve.

An *inferred mineral resource* is that part of a mineral resource for which quantity and grade or quality are estimated on the basis of limited geological evidence and sampling. Geological evidence is sufficient to imply, but not verify, geological and grade or quality continuity. An inferred resource is accompanied by a lower level of confidence than is an indicated mineral resource and must not be converted to a mineral reserve. It may be reasonably expected that the majority of inferred mineral resources could be upgraded to indicated mineral resources with continued exploration. An inferred mineral resource is accompanied by a lower level of confidence than is an indicated mineral resource.

Source: CRIRSCO 2013.

The Society of Petroleum Engineers–Petroleum Resources Management System (SPE-PRMS)

The SPE-PRMS is the main classification for the reporting of crude oil and natural gas. The SPE-PRMS is based on an explicit distinction between (i) the development project that has been (or will be) implemented to recover petroleum from one or more accumulations and, in particular, the chance of commerciality of that project; and (ii) the range of uncertainty in the petroleum quantities that are forecast to be produced and sold in the future from that development project.[3]

The SPE-PRMS (like the CRIRSCO) captures geological and economic factors within a two-dimensional framework: the vertical axis corresponds to the degree of commerciality of the resource, while the horizontal axis corresponds to its range of geological uncertainty. Three main categories are distinguished on the vertical axis: reserves, contingent resources, and prospective resources. Projects that are, for example, classified in the reserves category, satisfy all commerciality requirements. On the horizontal axis, at least three estimates of the geological certainty regarding the quantity to be extracted are captured. Depending on the degree of commerciality of the reserve/resource, these estimates are called proved, probable, and possible quantities—or low, best, and high estimates. Figure A.2 summarizes the main principles of the SPE-PRMS.

Figure A.2   SPE-PRMS Hydrocarbon Resources Classification Framework

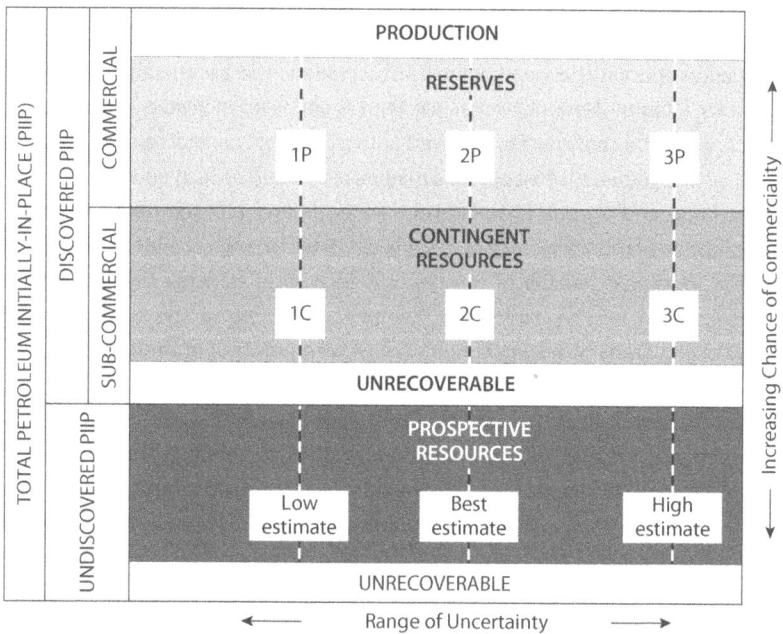

Source: Society of Petroleum Engineers (SPE) and others 2011.
Note: "Range of Uncertainty" reflects a range of estimated quantities potentially recoverable from an accumulation by a project. "Chance of Commerciality" reflects the chance that the project will be developed and will reach commercial producing status. 1P = proved reserves; 2P = proved plus probable reserves; 3P = proved plus probable plus possible reserves.

Box A.2 discusses the definitions corresponding to the subcategories of the SPE-PRMS framework in more detail.

The definitions following the box come from SPE and others (2011). Readers are encouraged to visit the SPE Web site, as definitions may change.

Box A.2 SPE-PRMS Classification Definitions

Reserves are those quantities of petroleum anticipated to be commercially recoverable by application of development projects to known accumulations from a given date forward under defined conditions. Reserves must further satisfy four criteria; that is, they must be (i) discovered, (ii) recoverable, (iii) commercial, and (iv) remaining (as of the evaluation date) based on the development project(s) applied. Reserves are further categorized in accordance with the level of certainty associated with the estimates and may be subclassified based on project maturity and/or characterized by development and production status:

- *Proved reserves* (1P) are those quantities of petroleum that, by analysis of geoscientific and engineering data, can be estimated with reasonable certainty to be commercially recoverable from a given date forward from known reservoirs and under defined economic conditions, operating methods, and government regulations. If deterministic methods are used, the term "reasonable certainty" is intended to express a high degree of confidence that the quantities will be recovered. If probabilistic methods are used, there should be at least a 90 percent probability that the quantities actually recovered will equal or exceed the estimate.
- *Probable reserves* are those additional reserves that analysis of geoscientific and engineering data indicate are less likely to be recovered than proved reserves, but more likely to be recovered than possible reserves. It is equally likely that actual remaining quantities recovered will be greater than, or less than, the sum of the estimated proved plus probable (2P) reserves. In this context, when probabilistic methods are used, there should be at least a 50 percent probability that the actual quantities recovered will equal or exceed the 2P estimate.
- *Possible reserves* are those additional reserves that analysis of geoscientific and engineering data indicate are less likely to be recoverable than probable reserves. The total quantities ultimately recovered from the project have a low probability to exceed the sum of proved plus probable plus possible (3P) reserves, which is equivalent to the high-estimate scenario. When probabilistic methods are used, there should be at least a 10 percent probability that the actual quantities recovered will equal or exceed the 3P estimate.

Contingent resources are those quantities of petroleum estimated, as of a given date, to be potentially recoverable from known accumulations, but the applied project(s) are not yet considered mature enough for commercial development due to one or more contingencies. Contingent resources may include, for example, projects for which there are currently no viable markets, or for which commercial recovery is dependent on technology under

---

box continues next page

Box A.2 SPE-PRMS Classification Definitions *(continued)*

development, or where evaluation of the accumulation is insufficient to clearly assess commerciality. Contingent resources are further categorized in accordance with the level of certainty associated with the estimates and may be subclassified based on project maturity and/or characterized by their economic status.

Prospective resources are those quantities of petroleum estimated, as of a given date, to be potentially recoverable from undiscovered accumulations by application of future development projects. Prospective resources have both an associated chance of discovery and a chance of development. Prospective resources are further subdivided in accordance with the level of certainty associated with recoverable estimates, assuming their discovery and development and may be subclassified based on project maturity.

Undiscovered petroleum initially-in-place is that quantity of petroleum estimated, as of a given date, to be contained within accumulations yet to be discovered.

Unrecoverable is that quantity of discovered or undiscovered petroleum initially-in-place that is estimated, as of a given date, not to be recoverable by future development projects. A portion of this quantity may become recoverable in the future as commercial circumstances change or technological developments occur; the remaining portion may never be recovered due to physical/chemical constraints such as those due to subsurface interaction of fluids and reservoir rocks.

Source: SPE and others 2011.

United Nations Framework Classification for Fossil Energy and Mineral Reserves and Resources 2009

The UNFC-2009 is a three-dimensional system that explicitly captures socioeconomic viability, in addition to geological and project feasibility. The UNFC-2009 is the third major system designed during the 1990s as a universally applicable scheme for classifying/evaluating energy and mineral reserves and resources.

Unlike the CRIRSCO and SPE-PRMS, this classification system is thought as an umbrella, relevant for both fossil energy and minerals. It is based on three dimensions. The first set of categories (the E axis) relates to the degree of favorability of social and economic conditions in establishing the commercial viability of the project, including consideration of market prices and relevant legal, regulatory, environmental, and contractual conditions. The second set (the F axis) designates the maturity of studies and commitments necessary to implement mining plans or development projects. These extend from early exploration efforts, before a deposit or accumulation has been confirmed to exist, to the extraction and sales of a commodity, and reflect standard value chain management principles. The third set of categories (the G axis) designates the level of confidence in the geological knowledge and potential recoverability of the quantities (UN 2010). Figure A.3 summarizes the main principles of the UNFC-2009 system.

Figure A.3 UNFC-2009 System: Key Principles

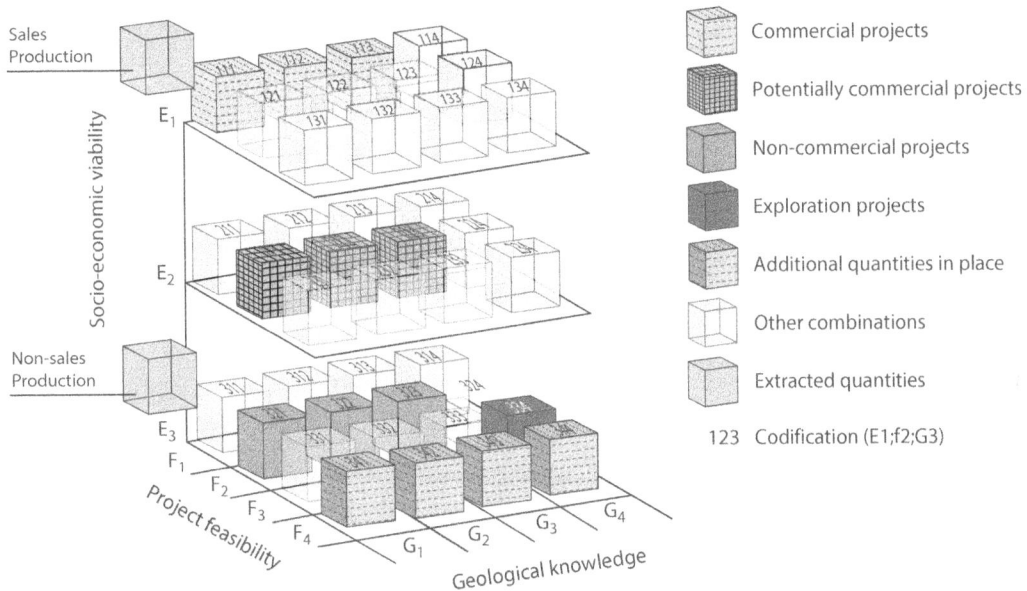

Source: United Nations Economic Commission for Europe (UN 2010).
Note: **E1**= Extraction and sale has been confirmed to be economically viable.
E2 = Extraction and sale is expected to become economically viable in the foreseeable future.
E3 = Extraction and sale is not expected to become economically viable in the foreseeable future or evaluation is at too early a stage to determine economic viability.
F1 = Feasibility of extraction by a defined development project or mining operation has been confirmed.
F2 = Feasibility of extraction by a defined development project or mining operation is subject to further evaluation.
F3 = Feasibility of extraction by a defined development project or mining operation cannot be evaluated due to limited technical data.
F4 = No development project or mining operation has been identified.
G1 = Quantities associated with a known deposit that can be estimated with a high level of confidence.
G2 = Quantities associated with a known deposit that can be estimated with a moderate level of confidence.
G3 = Quantities associated with a known deposit that can be estimated with a low level of confidence.

System of Environmental-Economic Accounting 2012

The SEEA-2012 sets parameters for the environmental accounting of renewable and nonrenewable natural resources and its integration into official economic statistics. The system introduces a new classification framework for the reporting of stocks of natural resources. Known deposits are categorized in three classes, defined by combinations of criteria from the UNFC-2009 system (see table A.1 for more on the categories listed):

- *Class A: Commercially recoverable resources.* This class includes deposits for projects that fall in categories E1 and F1 and where the level of confidence in the geologic knowledge is high (G1), moderate (G2), or low (G3).
- *Class B: Potentially commercially recoverable resources.* This class includes deposits for those projects that fall in category E2 (or eventually E1), and at

Table A.1 SEEA-2012 Classes and Relevant UNFC-2009 Categories

SEEA classes	Corresponding UNFC-2009 project categories		
	E	F	G
	Economic and social viability	Field project status and feasibility	Geologic knowledge
Known deposits			
A: Commercially recoverable resources	E1. Extraction and sale have been confirmed to be economically viable	F1. Feasibility of extraction by a defined development project or mining operation has been confirmed	Quantities associated with a known deposit that can be estimated with a high (G1), moderate (G2), or low (G3) level of confidence
B: Potential commercially recoverable resources	E2. Extraction and sale are expected to become economically viable in the foreseeable future	F2.1 Project activities are ongoing to justify development in the foreseeable future Or F2.2 Project activities are on hold and/or where justification as a commercial development may be subject to significant delay	
C: Noncommercial and other known deposits	E3. Extraction and sale are not expected to become economically viable in the foreseeable future or evaluation is at too early a stage to determine economic viability	F2.2 Project activities are on hold and/or where justification as a commercial development may be subject to significant delay Or F2.3 There are no current plans to develop or to acquire additional data at the time due to limited potential Or F4. No development project or mining operation has been identified	
Potential deposits (not included in SEEA) Exploration projects Additional quantities in place	E3. Extraction and sale are not expected to become economically viable in the foreseeable future or evaluation is at too early a stage to determine economic viability	F3. Feasibility of extraction by a defined development project or mining operation cannot be evaluated due to limited technical data Or F4. No development project or mining operation has been identified	Estimated quantities associated with a potential deposit, based primarily on indirect evidence (G4)

Source: Adapted from OECD (2014).

the same time in F2.1 to F2.2, and where the level of confidence in the geologic knowledge is high (G1), moderate (G2), or low (G3).

- *Class C: Noncommercial and other known deposits.* These are resources for those projects that fall into category E3, and for which the feasibility is categorized as F2.2, F2.3, or F4 and where the level of confidence in the geologic knowledge is high (G1), moderate (G2), or low (G3).

Deposits exclude potential ones for which there is no expectation of becoming economically viable, inadequate information to determine the feasibility of extraction, or insufficient confidence in the geologic knowledge. Table A.1 provides an overview of how the classes of resources are defined, based on the UNFC-2009 criteria.

Types of Economic Rents

Hotelling Rents, or User Costs

In 1931, Harold Hotelling, an American economist, published a seminal article titled "The Economics of Exhaustible Resources." The article pointed out that firms exploiting a nonrenewable resource behave differently from firms that depend on renewable or unlimited resources. Hotelling noted that firms using nonrenewable resources incur an opportunity cost, in addition to their production costs, in the process of producing finite mineral or hydrocarbon commodities. This is because increasing output by one more unit today, rather than leaving the required mineral resources in the ground, reduces the mineral resources available in the future (Otto and others 2006).

More specifically, the opportunity cost, or "Hotelling rent" or "user cost" is the net present value (NPV) of the future profits that are lost because mineral resources are reduced by an additional unit of output today. With this framework in mind, profit-maximizing, competitive firms producing mineral commodities will only expand their output up to the point at which the market price equals the production costs of the last unit plus its opportunity cost. Otherwise, the firm's profitability is enhanced by ceasing production today, and saving its mineral resources for the future.

Hotelling rent is not really a "rent," and its taxation is distortionary. Some authors prefer the term "user cost" to "Hotelling rent," because if the market price does not cover this opportunity cost plus the current costs of production, a profit-maximizing producer will be incentivized to shut down and leave its resources in the ground for depletion at a time in the future. Thus, the user cost (or Hotelling rent, or scarcity rent) reflects real costs and is not really an economic rent at all. As a result, its collection through taxation is distortionary in the short and long run, in the sense that it alters optimal economic behavior and the efficient allocation of resources (figure 2.3 in the main text).

Despite being of great interest to the academic community, and included in theoretical models of depletion, Hotelling rents tend to be very small in comparison to Ricardian or quasi-rents, if present at all (see Halvorsen and Smith

[1991], for example). Furthermore, Hotelling rents are generally ignored in industry decisions—that is, in depletion planning processes. As a result, rents are principally believed to be of the Ricardian type, though in the case of some markets (oil, potash/phosphate, and, until recently, diamonds), monopoly/oligopoly rents can exist, and these can be material.

Ricardian Rents

The classical theory of resource rents is associated with David Ricardo, a British economist and one of the first to explore economic rents. Writing in the early nineteenth century, he noted that agricultural land could be categorized according to its fertility: land in the most fertile class can produce a given quantity of food at lower cost than land in the second-most fertile class. Similarly, land in the second-most fertile class has costs below that of land in the third class, and so on. Assuming that all other factors are equal, due to the differential productivity inherent in the land itself (as opposed to the skill of the farmer), the owner of the land could charge a higher rent to the farmer who used the first class of land than he could to one who used the second or third classes of land.

Similarly, in the extractive industries, Ricardian rent is generally defined as a differential rent due to the heterogeneous nature of subsoil assets. Some subsoil assets are of higher quality than others, and these quality advantages are reflected in lower costs of production, higher prices for the products these assets yield, or both. A Ricardian rent may reflect, for example, large size, high grade, ease of processing, or good location.

In theory, Ricardian rents can be fully taxed away without affecting decision making; however, in many cases economists and governments incorrectly commingle Ricardian rents with Hotelling and quasi-rents. The latter only appear as rents because in practice we observe annual profits from the extractives companies but ignore historical investments in discovery and construction of facilities. In Ricardo's *On the Principles of Political Economy and Taxation* (1821), he defines rent rigorously and touches on the notion of quasi-rents (see box B.1).

Quasi-Rents

Quasi-rents exist only in the short run and reflect returns to capital and other fixed costs. In the long run, a mine or oil and gas producer not recovering its fixed costs will shut down.

Returns due to imperfect markets can be grouped as quasi-rents because of their often transient nature. Monopoly, oligopoly, or monopsony rents are the portion of overall value due to the market power of one or a few dominant sellers or buyers, and their ability to raise (or lower, in the case of monopsony) the market price above what would exist in a competitive market. Oligopoly rents could occur, for example, in the oil sector through the efforts of the Organization

Box B.1   Rents Outlined in David Ricardo's 1821 Treatise *On the Principles of Political Economy and Taxation*

"Rent is that portion of the produce of the earth, which is paid to the landlord for the use of the original and indestructible powers of the soil…. It is often, however, confounded with the interest and profit of capital, and, in popular language, the term is applied to whatever is annually paid by a farmer to his landlord. If, of two adjoining farms of the same extent, and of the same natural fertility, one had all the conveniences of farming buildings, and, besides, were properly drained and manured, and advantageously divided by hedges, fences and walls, while the other had none of these advantages, more remuneration would naturally be paid for the use of one, than for the use of the other; yet in both cases this remuneration would be called rent. But it is evident, that a portion only of the money annually to be paid for the improved farm, would be given for the original and indestructible powers of the soil; the other portion would be paid for the use of the capital which had been employed in ameliorating the quality of the land, and in erecting such buildings as were necessary to secure and preserve the produce."

On the rent of mines, Ricardo writes:

"The metals, like other things, are obtained by labour. Nature, indeed, produces them; but it is the labour of man which extracts them from the bowels of the earth, and prepares them for our service.

"Mines, as well as land, generally pay a rent to their owner; and this rent, as well as the rent of land, is the effect, and never the cause of the high value of their produce.

"If there were abundance of equally fertile mines, which any one might appropriate, they could yield no rent; the value of their produce would depend on the quantity of labour necessary to extract the metal from the mine and bring it to market.

"The metal produced from the poorest mine that is worked, must at least have an exchangeable value … sufficient to procure all [inputs consumed in working it]. The return for capital from the poorest mine paying no rent, would regulate the rent of all the other more productive mines. This mine is supposed to yield the usual profits of stock. All that the other mines produce more than this, will necessarily be paid to the owners for rent. Since this principle is precisely the same as that which we have already laid down respecting land, it will not be necessary further to enlarge on it."

Source: Ricardo 1821.

of the Petroleum Exporting Countries (OPEC). Monopsony rents may materialize where a mine finds itself the only employer in the region and therefore is able to purchase labor inputs at rates lower than otherwise possible. Smelters and refiners, because of the large economies of scale required, can also generate monopsony rents.

APPENDIX C

Impact of Income Changes on Commodity Demand

How Does Demand for Commodities Adjust?

The demand for commodities is responsive to short-run income changes caused by business-cycle fluctuations. Metals and other materials are consumed primarily in the capital equipment, construction, transportation, and consumer durables sectors of the economy. These sectors boom when the economy is doing well, and spending on these goods is significantly curtailed when the economy is in a downturn.

Commodity demand also adjusts over the long run as a response to long-run structural changes in the economy. The gradual structural transformation of an economy from agriculture toward services, as illustrated through stylized "Kuznets facts" (figure C.1), give rise to the inverted-U-shaped "intensity of use" (IoU) relationships (commodity demand as a proportion of gross domestic product, GDP) described in figures C.1.1 and C.1.2. This is driven by the fact that, as countries begin to industrialize, they invest in infrastructure, manufacturing, and processing industries—as in the case of China, where much of the increased consumption has been driven by large-scale urbanization and resulting IoU increases (Malenbaum 1975, 1978). But as the economy continues to grow, demand for "industrialization" materials begins to wane and is replaced by maintenance of the stock of human-made physical assets—and economic growth becomes more intensively driven by services.

The IoU technique is a simple procedure that can be used to assess changes over time, and between countries, in metal or energy product demand. Despite its ease of adoption, the technique suffers from some shortcomings, however. The most significant of these is the implicit assumption that IoU depends solely on per capita income, and thus the relationship between the two is linear. As a result, material price changes, new technologies, and other factors whose influence is unlikely to vary in line with per capita income also affect IoU. This means that the true inverted-U-shaped relationship between IoU and per capita income shifts downward over time, due to new resource-saving technology. But material substitution and new technology occasionally push the IoU curve upward. Box C.1 provides an example of the IoU approach, applied to China.

Figure C.1   The "Kuznets Facts," Illustrated by the Share of U.S. Employment in Agriculture, Manufacturing, and Services, 1800–2000

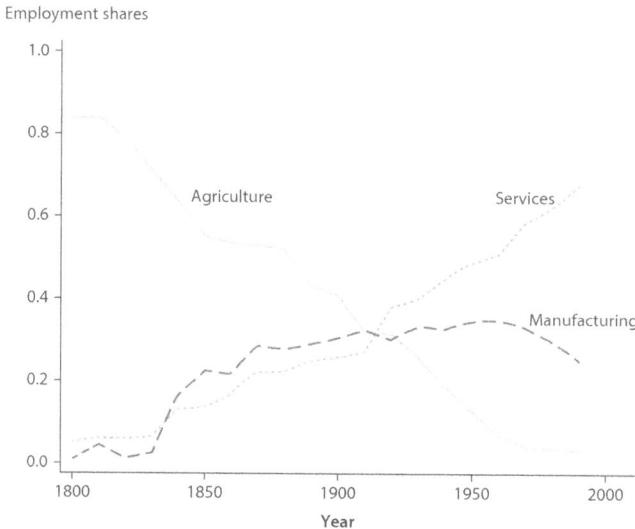

Source: Acemoglu 2008. Reprinted with permission of Princeton University Press; permission conveyed through the Copyright Clearance Center.

Box C.1   Secular Growth and Structural Change in China: An Application of the Intensity-of-Use Approach

Like many extractive industries (EI) companies, Anglo American plc utilizes the intensity-of-use (IoU) approach to better anticipate changes to long-term demand for the commodities it produces. As can be seen in figure C.1.1, the IoU plots economy activity per capita (that is, GDP/capita) against demand for a particular commodity per unit of GDP. This representation is useful as it allows demand for mineral commodities to be projected using national growth forecasts.

One of the world's largest mining companies, Anglo, notes that as China continues to urbanize and industrialize, and shifts from an investment-intensive to a consumption-driven economy, the growth rate in demand for certain commodities such as steel materials (that is, iron ore, ferrochrome, and so on) is expected to moderate.

This shift to increased consumption relative to investment is, however, expected to drive a stronger demand growth rate for more luxury or discretionary commodities such as diamonds (driven by increased demand for luxury products) and platinum group metals (driven by increased demand for automobiles).

Despite being intellectually appealing, the IoU concept has its limitations. For example, every country is unique, and so, although there may be general trends, one country's material intensity will not be exactly the same as another's. This is clearly seen in figure C.1.2, where

box continues next page

Box C.1 Secular Growth and Structural Change in China: An Application of the Intensity-of-Use Approach (continued)

China's current steel intensity is shown relative to those of other developed countries. The IoU is therefore unique not only depending on the commodity in question, but also to the country, making accurate projections from its use difficult.

Figure C.1.1 Indexed Intensity of Use in China for Various Commodities

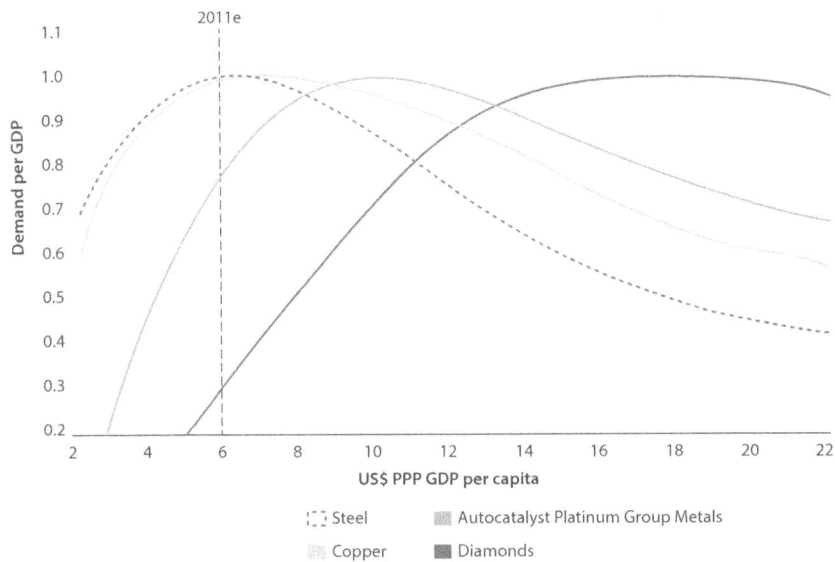

Source: Anglo American plc 2012.

Figure C.1.2 Steel Intensity and Gross Domestic Product in Selected Countries, 1900–2011
kg/capita crude steel production

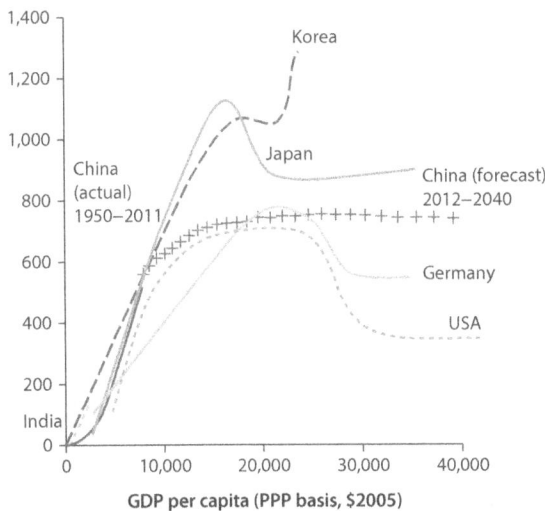

Source: Rio Tinto 2013.
Note: GDP = gross domestic product; PPP = purchasing power parity.

Notes

1. The Committee for Mineral Reserves International Reporting Standards (CRIRSCO) was formed in 1994 under the auspices of the Council of Mining and Metallurgical Institutes (CMMI). It is a grouping of representatives of organizations that are responsible for developing mineral reporting codes in Australasia (Joint Ore Reserves Committee, JORC), Canada (Canadian Institute of Mining, Metallurgy and Petroleum, CIM), Chile (National Committee), Europe (National Committee Pan-European Reserves and Resources Reporting, PERC), Mongolia (Mongolian Professional Institute of Geoscience and Mining, MPIGM), Russia (NAEN), South Africa (South African Code for Reporting of Exploration Results, Mineral Resources, and Reserves, SAMREC), and the United States (Society for Mining, Metallurgy, and Exploration).

2. Care should always be taken when reading resource and reserve statements, which may list resources inclusive or exclusive of reserves. For example, according to the Canadian reporting standard: "When reporting both Mineral Resources and Mineral Reserves, a clarifying statement must be included that clearly indicates whether Mineral Reserves are part of the Mineral Resource or that they have been removed from the Mineral Resource. A single form of reporting should be used in a report" (CIM 2014).

3. http://www.spe.org/industry/docs/PRMS_Guidelines_Nov2011.pdf, p. 7.

Effective Resource Contract Enforcement: A Checklist of Guidelines

Why Use a Checklist?

The following set of questions are a potential guide for ministry of finance officials, to ensure that the ministry is aware of some of the critical issues in resource contract management that may impact on budget planning and budget execution. (This list of questions is not exhaustive, and the questions are not in order of importance. Specific country conditions and particular mining projects may imply that other issues not mentioned here become salient in resource contract enforcement.)

Resource Revenue Collection

- *Database for resource revenue obligations.* Is there an integrated database of all the terms of the fiscal obligations—tax rates, royalty rates, specified signing and production bonuses—associated with a particular mining/petroleum project? Who is responsible for maintaining this database, such that any renegotiation of contractual terms is reflected in a timely manner?
- *Verification of production and export figures.* Are the revenue-collecting authority's production and export figures for the resource verified with the relevant technical agency responsible for the monitoring of operational and production commitments? Are there any discrepancies between the two, and are they reconciled on a regular basis?
- *Verification of quality of resource exported.* Commodity prices and, subsequently, company profits can differ significantly according to the quality of the commodity exported (for example, different ore grades). The quality of commodity exports thus needs to be verified and monitored by the ministry of mineral resources (or other relevant technical agency). Has the revenue-collecting authority coordinated with the relevant technical agency to ensure that the commodity price underlying the calculation of profit-based taxes is consistent with the quality of the commodity exported?

- *Analysis of historical resource revenue collection.* Have historical resource revenue collection figures been analyzed for potential gaps in revenue collection efforts, such that revenue streams that have shown persistent weaknesses (that is, areas that have exhibited substantial shortfall in collection as against projection) can be identified? Is there a systematic approach to addressing these areas of persistent weaknesses in collection?
- *Resource revenue flow of funds.* Is there a clear understanding of the flow of funds for resource revenues, and are the various resource revenue accounts reconciled on a regular basis? For instance, is all the revenue transferred to a single treasury account (consolidated fund) by the main revenue-collecting agency? Or, alternatively, are individual ministries, departments, or agencies (MDAs) responsible for collecting particular revenue streams, and do these MDAs collect revenue in a commercial bank and then transfer funds to the treasury?

Resource Revenue Projections and Macrofiscal Planning

- *Verification of commodity price assumption.* Are the commodity price assumptions used for medium-term projections of resource revenue realistic, transparent, and made in conjunction with the ministry of mineral resources and other relevant technical ministries (see Ossowski and Halland [forthcoming, 2015], vol. 2 of the *Essentials for Economists, Public Finance Professionals, and Policy Makers,* "Fiscal Management in Resource-Rich Countries," for a detailed discussion of commodity prices to be used for budget purposes).
- *Verification of production and export forecasts.* Have the production and export forecasts for commodities used in the macrofiscal framework, and the national account forecasts, been verified with the technical agencies involved, such as the ministry for mineral resources?
- *Incorporation of production and commodity price risks.* Has the medium-term macrofiscal framework incorporated risks to resource revenue collection (driven by operational risks inherent in production and export), as well as volatility in global commodity prices? For instance, does the budget framework paper contain an analysis of the potential strategic budgetary choices relevant to expenditure reduction or additional borrowing in response to a given fall in commodity prices below forecast? (See Ossowski and Halland [forthcoming, 2015] for details.)
- *Life cycle of an oil or mining project.* Does the medium-term resource revenue forecast take into account the typical life cycle of an oil or mining project? For instance, revenues from royalties typically remain steady throughout the life of a project, but profit-based taxes could become significant only once all capital expenditures have been amortized, which may be several years into the life of a project.
- *In-year timing of revenue receipts.* Has the in-year timing of the revenue receipts from the mine been identified? This may be especially important to do in resource-revenue-dependent countries with thin domestic financial markets

and a limited ability to borrow externally. In these countries, to ensure effective liquidity management, expenditure sequencing, and/or short-term borrowing strategy within a year, the timing of revenue receipts may have to be taken into account.

- *Foreign exchange and balance of payments.* To aid macrofiscal planning and to forecast the balance of payments, has the extent of foreign exchange in resource revenue transactions, as well as in the potential investments made in the country by companies prior to production, been ascertained?

Management of Expenditure and Contingent Liabilities

- *Revenue share and payments.* Some resource contracts specify a certain percentage or amount of the resource revenue that needs to be transferred to particular regions or subnational governments or to particular government MDAs. Have these liabilities been identified and verified with the relevant subnational governments or government MDAs, and has this information been duly incorporated into the medium-term expenditure framework (MTEF)? Is there an understanding that volatility in resource revenue collection may lead to additional expenditure pressures for the central budget?
- *Community development fund projects.* Resource contracts at times specify the creation of community development funds, with company support, that may include projects in local communities (see chapter 6 of this volume, the section on community foundations, trusts, and funds). Are any such projects consistent with the broader public investment program for the region, and for the country? Furthermore, have all the central government funding requirements, in terms of counterpart funding and recurrent maintenance costs, for these projects been accurately identified?
- *Resources for infrastructure and recurrent costs.* Resource extraction projects often involve building ancillary infrastructure, and resource contracts may specify that infrastructure be built in exchange for resource extraction. The implications of a "resource-financed infrastructure" (RFI) deal and its potential relevance to public financial management (PFM) are discussed in further detail in the main text of this volume (see chapter 7, box 7.2). It should be noted that fiscal space must be created in the medium term for the recurrent costs of maintaining infrastructure.
- *Costs of displacement and resettlement.* Resource contracts may specify specific terms for the resettlement and rehabilitation of communities directly affected by resource extraction. Have the companies adhered to these terms, and does the government have any potential contingent liability for the cost of resettlement?
- *Identifying contingent environmental liabilities.* Have the environmental management plans (EMPs) submitted by companies been analyzed to identify potential liabilities that may arise due to environmental damage related to

resource extraction? Have the risks associated with environmental damage and the potential extent of damage been evaluated in conjunction with technical departments responsible for environmental management (such as the ministry of environment)? Are the established financial surety arrangements sufficient to cover potential liabilities from decommissioning?

References

Acemoglu, Daron. 2008. "Structural Change and Economic Growth." In *Introduction to Modern Economic Growth*. Princeton, NJ: Princeton University Press.

Acemoglu, Daron, and James A. Robinson. 2006. "Economic Backwardness in Political Perspective." *American Political Science Review* 100 (February): 115–31.

Agosin, Manuel R., Christian Larrain, and Nicolas Grau. 2010. "Industrial Policy in Chile." IDB Working Paper Series No. IDB-WP-170, Inter-American Development Bank, Washington, DC.

Aguilar, Javier, and Greg Francis. 2005. "Cajamarca Mining Canon Project, Phase 1: Promoting Public-Private Partnerships in a Context of Mistrust." Unpublished note, International Finance Corporation, Washington, DC.

Ahmad, Ehtisham, and Eric Mottu. 2003. "Oil Revenue Assignment: Country Experiences and Issues." Chapter 9 in *Fiscal Policy Formulation and Implementation in Oil-Producing Countries*, ed. J. M. Davis, R. Ossowski, and A. Fedelino. Washington, DC: IMF.

Alexeev, Michael, and Robert Conrad. 2009. "The Elusive Curse of Oil." *The Review of Economics and Statistics* 91 (3): 589–98.

Anderson, George, ed. 2012. *Oil and Gas in Federal Systems*. London and New York: Oxford University Press.

Anglo American plc. 2012. *Anglo American plc. 2011 Annual Report*. London, England. http://www.angloamerican.com.

Ascher, William. 2008. Presentation at the UN Research Institute for Social Development workshop on "Financing Social Policy in Mineral-Rich Countries," Geneva, April 24–25, 2008.

Auditor General of South Africa. 2009. *Report of the Auditor General to the Parliament on a Performance Audit of the Rehabilitation of Abandoned Mines at the Department of Minerals and Energy, South Africa*. Auditor General of South Africa, Pretoria.

Baer, Katherine. 2002. "Improving Large Taxpayer's Compliance. A Review of Country Experience." IMF Occasional Paper 215, International Monetary Fund, Washington, DC.

Bawumia, Mahamudu, and Håvard Halland. Forthcoming, 2015. "From Boom to Gloom: Ghana's Oil Discovery and Its Effects on Fiscal Variables." World Bank Policy Research Working Paper, Washington, DC.

Benavente, Jose. 2006. "Wine Production in Chile." In *Technology, Adaptation, and Exports: How Some Developing Countries Got It Right*, ed. Vandana Chandra. Washington, DC: World Bank.

Berg, Andrew, Rafael Portillo, Shu-Chun S. Yang, and Luis-Felipe Zanna. 2012. "Public Investment in Resource-Abundant Developing Countries." IMF Working Paper 12/274, International Monetary Fund, Washington, DC.

BGS International. 2012. "Geodata for Development, A Practical Approach." In *EI Source Book*, ed. Peter Cameron and Michael Stanley. http://www.eisourcebook.org/.

Bleaney, Michael, and Håvard Halland. 2014. "Natural Resource Exports, Fiscal Policy Volatility and Growth." *Scottish Journal of Political Economy* 61 (5): 502–22.

———. Forthcoming, 2015. "Do Resource-Rich Countries Suffer from a Lack of Fiscal Discipline?" World Bank Working Paper, Washington, DC.

Bourguignon, François, and Thierry Verdier. 2000. "Oligarchy, Democracy, Inequality, and Growth." *Journal of Development Economics* 62: 285–313.

BP. 2008. "BP Statistical Review of World Energy." http://www.bp.com/liveassets/bp_internet/globalbp/globalbp_uk_english/reports_and_publications/statistical_energy_review_2008/STAGING/local_assets/downloads/pdf/statistical_review_of_world_energy_full_review_2008.pdf.

Brosio, Giorgio. 2003 "Oil Revenue and Fiscal Federalism." Chapter 10 in *Fiscal Policy Formulation and Implementation in Oil-Producing Countries*, ed. J. M. Davis, R. Ossowski, and A. Fedelino. Washington, DC: IMF.

Brunnschweiler, Christa N., and Erwin H. Bulte. 2006. "The Resource Curse Revisited and Revised: A Tale of Paradoxes and Red Herrings." CER-ETH Economics Working Paper Series 06/61, Center of Economic Research, ETH Zurich. https://ideas.repec.org/p/eth/wpswif/06-61.html.

Calder, Jack. 2010. "Resource Tax Administration: Functions, Procedures, and Institutions." In *The Taxation of Petroleum and Minerals, Principles, Problems, and Practice*, ed. Philip Daniel, Michael Keen, and Charles McPherson. London and New York: Routledge.

———. 2014. *Administering Fiscal Regimes for the Extractive Industries: A Handbook*. Washington, DC: International Monetary Fund. http://www.elibrary.imf.org/doc/IMF071/20884-9781475575170/20884-9781475575170/Other_formats/Source_PDF/20884-9781484386446.pdf.

Cameron, Peter, and Michael Stanley. 2012. *EI Source Book*. Oil, Gas, and Mining Policy Unit (SEGOM), World Bank Group; and Centre for Energy, Petroleum, and Mineral Law and Policy, University of Dundee. http://www.eisourcebook.org.

Central Bank of Russia. 2011. "Production Sharing Agreements." Paper prepared for the 24th meeting of the IMF Committee on Balance of Payments Statistics, Central Bank of Russia, Moscow, October 24–26.

CIM (Canadian Institute of Mining, Metallurgy, and Petroleum). 2014. "CIM Definition Standards—For Mineral Resources and Mineral Reserves." Document prepared by the CIM Standing Committee on Reserve Definitions, adopted by the CIM Council on May 10, 2014. http://www.cim.org/~/media/Files/PDF/Subsites/CIM_DEFINITION_STANDARDS_20142.

Collier, Paul, and Anke Hoeffler. 2004. "Greed and Grievance in Civil War." *Oxford Economic Papers* 56: 563–95.

Corden, Max, and Peter Neary. 1982. "Booming Sector and De-industrialisation in a Small Open Economy." *The Economic Journal* 92 (December): 825–48.

CRIRSCO (Committee for Mineral Reserves International Reporting Standards). 2013. "International Reporting Template for the Public Reporting of Exploration Results,

Mineral Resources, and Mineral Reserves." http://www.crirsco.com/templates/crirsco_international_reporting_template_2013.pdf.

———. 2015. "About CRIRSCO." Committee for Mineral Reserves International Reporting Standards—Background (accessed March 20, 2015). http://www.crirsco.com/background.asp.

Dabla-Norris, Era, Jim Brumby, Annette Kyobe, Zac Mills, and Chris Papageorgiou. 2011. "Investing in Public Investment: An Index of Public Investment Efficiency." IMF Working Paper 11/37, International Monetary Fund, Washington, DC.

Daniel, Philip, Michael Keen, and Charles McPherson, eds. 2010. *The Taxation of Petroleum and Minerals, Principles, Problems, and Practice.* London and New York: Routledge.

Davis, Graham A., and John E. Tilton. 2005. "The Resource Curse." *Natural Resources Forum* 29 (3): 233–42. http://lawweb.colorado.edu/profiles/syllabi/banks/Davis%2520%2520Tilton%2520-%2520The%2520resource%2520curse.pdf.

Department of Energy and Climate Change, United Kingdom. 2011. "Guidance Notes for Industry on the Decommissioning of Offshore Oil and Gas Installations and Pipelines under the Petroleum Act, 1998." https://www.gov.uk/government/uploads/system/uploads/attachment_data/file/69754/Guidance_Notes_v6_07.01.2013.pdf.

Eggert, Roderick. 2011. "Critical Elements and Thin-Film Photovoltaics." Unpublished presentation presented at the National Renewable Energy Laboratory, Golden, Colorado, October 14.

EITI (Extractive Industries Transparency Initiative). 2013. "The EITI Standard." http://eiti.org/document/standard.

Frankel, Jeffrey A. 2010. "The Natural Resource Curse: A Survey." NBER Working Paper 15836, National Bureau of Economic Research, Cambridge, MA.

Gammon, John B. 2007. "A Draft Mineral Sector Policy Note for Serbia." World Bank, Washington, DC, unpublished.

Gelb, Alan. 2011. "Economic Diversification in Resource Rich Countries." In *Beyond the Curse: Policies to Harness the Power of Natural Resources,* ed. Rabah Arezki, Thorvaldur Gylfason, and Amadou Sy. Washington, DC: International Monetary Fund.

Ghana Public Interest and Accountability Committee. 2012. "Ghana Public Interest and Accountability Committee—Annual Report 2011." Ghana Public Interest and Accountability Committee, Accra, Ghana.

Government of Afghanistan. 2010. "National Extractive Industry Excellence Program." Ministry of Mines, Government of Afghanistan, Kabul. http://www.acbar.info/uploads/Publication/02_NEIEP_project_document_revised_29March2011.pdf.

Government of Canada. 2006. "Mining Kit for Aboriginal Communities: Mining Sequence." Since replaced by the *Exploration and Mining Guide for Aboriginal Communities.* http://www.nrcan.gc.ca/mining-materials/aboriginal/7819.

Government of Indonesia. 2014. "Implementatiasi UU RI Nomor 4 Tahum 2009 Dan DampaknyaTerhadap Kebijakan Hilirisasi Pertambagan Mineral Dan Batubara." Presentation made by the directorate-general, Mineral and Coal, Government of Indonesia, at a Ministry of Trade dissemination event, February.

Government of New South Wales, Department of Industry and Investment. 2010. "Rehabilitation Cost Estimate Guidelines." New South Wales, Australia. http://cer.org.za/wp-content/uploads/2011/10/AG_Report_on_abandoned_mines-Oct-2009.pdf.

———. 2012. "ESG2: Environmental Impact Assessment Guidelines." Mineral Resources Environmental Sustainability Unit, Maitland, Australia.

Government of Uganda. 2008. "National Oil and Gas Policy for Uganda." Ministry of Energy and Mineral Development, Government of Uganda, Kampala. http://www.acode-u.org/documents/oildocs/oil&gas_policy.pdf.

Guj, Pietro, Boubacar Bocoum, James Limerick, Murray Meaton, and Bryan Maybee. 2013. *How to Improve Mining Tax Administration and Collection Frameworks: A Sourcebook.* Washington, DC: World Bank. http://www-wds.worldbank.org/external/default/WDSContentServer/WDSP/IB/2013/10/11/000442464_20131011125523/Rendered/PDF/818080WP0P12250Box0379844B00PUBLIC0.pdf.

Gupta, Sanjeev, Alvar Kangur, Chris Papageorgiou, and Abdoul Wane. 2011. "Efficiency-Adjusted Public Capital and Growth." IMF Working Paper 11/217, International Monetary Fund, Washington, DC. http://www.imf.org/external/pubs/ft/wp/2011/wp11217.pdf.

Gylfason, Thorvaldur, Tryggvi Thor Herbertsson, and Gylfi Zoega. 1999. "A Mixed Blessing: Natural Resources and Economic Growth." *Macroeconomic Dynamics* 3: 205–25.

Halland, Håvard, John Beardsworth, Bryan Land, and James Schmidt. 2014. "Resource Financed Infrastructure: A Discussion on a New Form of Infrastructure Financing." World Bank Studies Series. http://www-wds.worldbank.org/external/default/WDSContentServer/WDSP/IB/2014/06/06/000333037_20140606143941/Rendered/PDF/884850PUB0Box300EPI2102390May292014.pdf.

Halvorsen, Robert, and Tim R. Smith. 1991. "A Test of the Theory of Exhaustible Resources." *The Quarterly Journal of Economics* 106 (1): 123–40.

Hausman, Ricardo, Bailey Klinger, and Robert Lawrence. 2007. "Examining Beneficiation." Working Paper, Center for International Development, Kennedy School of Government, Harvard University.

Heller, Patrick, Paasha Mahdavi, and Johannes Schreuder. 2014. "Reforming National Oil Companies: Nine Recommendations." Natural Resource Governance Institute. http://www.resourcegovernance.org/publications/reforming-national-oil-companies-nine-recommendations.

Heum, Per. 2008. "Local Content Development—Experiences from Oil and Gas Activities in Norway." SNF Working Paper 02/08, Bergen, Norway.

Hotelling, Harold. 1931. "The Economics of Exhaustible Resources." *Journal of Political Economy* 39 (2): 137–75.

ICMM (International Council on Mining and Metals). 2013. "Approaches to Understanding Development Outcomes from Mining." ICMM report, International Council on Mining and Metals, London. http://www.icmm.com/document/5774.

IFC (International Finance Corporation). 2013. "Fostering the Development of Greenfield Mining-Related Transport Infrastructure through Project Financing." IFC, Washington, DC. http://www.ifc.org/wps/wcm/connect/c019bf004f4c6ebfbd99ff032730e94e/Mine+Infra+Report+Final+Copy.pdf?MOD=AJPERES.

IMF (International Monetary Fund). 2007. *Guide on Resource Revenue Transparency.* Washington, DC: IMF.

———. 2012. *Macroeconomic Policy Frameworks for Resource-Rich Developing Countries.* Washington, DC: IMF.

International Bar Association. 2011. "Model Mining Development Agreement Project." http://www.mmdaproject.org/.

Johnson, Robert C., and Guillermo Noguera. 2012. "Fragmentation of Trade in Value Added Over Four Decades." NBER Working Paper 18186, National Bureau of Economic Research, Cambridge, MA.

Jourdan, Paul. 2014. "Ownership and Mineral-based Development: The Role of State Institutions in the Minerals and Energy Sector." Presentation given at the Botswana Confederation of Commerce, Industry and Manpower Workshop, Maun, November.

Kaneva, Natasha. 2014. *Metals Outlook: Exploring Bull and Bear Risk Factors*. J.P. Morgan Commodities Research.

Kapstein, Ethan, and Rene Kim. 2011. *The Socio-Economic Impact of Newmont Ghana Gold Limited*. Accra: Stratcomm Africa. http://www.newmont.com/files/doc_downloads/africa/ahafo/environmental/Socio_Economic_Impact_of_Newmont_Ghana_Gold_July_2011_0_0.pdf.

Land, Bryan. 2010. "Resource Rent Taxes: A Re-appraisal." In *The Taxation of Petroleum and Minerals, Principles, Problems, and Practice*, ed. Philip Daniel, Michael Keen, and Charles McPherson. London and New York: Routledge.

Londono, David, and Benjamin Sanfurgo. 2014. "Technical Report on the Lumwana Mine, North-Western Province, Republic of Zambia." NI 43–101 Technical Report, Barrick Gold Corporation, Toronto, Canada.

Malenbaum, Wilfred. 1975. "Law of Demand for Minerals." *Proceedings of the Council of Economics*, 104th AIME Annual Meeting, New York.

———. 1978. *World Demand for Raw Materials in 1985 and 2000*. New York: McGraw Hill.

Matsuyama, K. 1992. "Agricultural Productivity, Comparative Advantage, and Economic Growth." *Journal of Economic Theory* 58: 317–443.

Mayorga Alba, Eleodoro. 2009. "Extractive Industries Value Chain: A Comprehensive Integrated Approach to Developing Extractive Industries." Africa Region Working Paper Series #125, Extractive Industries for Development Series #3, World Bank, Washington, DC.

McLure, Charles E. 2003 "The Assignment of Oil Tax Revenue." Chapter 8 in *Fiscal Policy Formulation and Implementation in Oil-Producing Countries*, ed. J. M. Davis, R. Ossowski, and A. Fedelino. Washington, DC: IMF.

McPherson, Charles. 2010. "State Participation in the Natural Resource Sectors: Evolution, Issues, and Outlook." In *The Taxation of Petroleum and Minerals, Principles, Problems, and Practice*, ed. Philip Daniel, Michael Keen, and Charles McPherson. London and New York: Routledge.

Mehlum, Halvor, Karl Moene, and Ragnar Torvik. 2006. "Institutions and the Resource Curse." *The Economic Journal* 116: 1–20.

Miller, Robert A. 2000. "Ten Cheaper Spades: Production Theory and Cost Curves in the Short Run." *Journal of Economic Education* (Spring): 119–30.

Minerals Council of Australia. 2015. "Life Cycle of a Mine" (accessed March 20, 2015). http://www.minerals.org.au/resources/gold/life_cycle_of_a_mine.

MinEx Consulting Pty Ltd. 2013. "Long Term Outlook for the Global Exploration Industry—Gloom or Boom?" Presentation by Richard Schodde to the Geological

Society of South Africa—GeoForum 2013 Conference, Johannesburg, South Africa, February 5.

MonTec. 2007. "Guidelines on Financial Guarantees and Inspections for Mining Waste Facilities." European Commission, Director General for Environment. http://ec. europa.eu/environment/waste/mining/pdf/EU_Final_Report_30.04.08.pdf.

Mullins, Peter. 2010. "International Tax Issues for the Resources Sector." In *The Taxation of Petroleum and Minerals, Principles, Problems, and Practice*, ed. Philip Daniel, Michael Keen, and Charles McPherson. London and New York: Routledge.

Nair, Arvind, and Yue Man Lee. 2014. "A Closer Look at Indonesia's Unprocessed Mineral Export Ban." *The Indonesia Economic Quarterly*, March 2014, World Bank, Washington, DC.

Nakhle, Carloe. 2010. "Petroleum Fiscal Regimes: Evolution and Challenges." In *The Taxation of Petroleum and Minerals, Principles, Problems, and Practice*, ed. Philip Daniel, Michael Keen, and Charles McPherson. London and New York: Routledge.

Natural Resources Canada. 2012. "Mining Information Kit for Aboriginal Communities." http://www.pdac.ca/pdac/advocacy/aboriginal-affairs/2006-mining-toolkit-eng.pdf.

———. 2013. "Feasibility Studies Table." Minerals and Metals Sector, Natural Resources Canada.

Newmont. 2013. "Mining 101: Understanding the Different Phases of Mine Operations." Blog entry, *Our Voice—Blog*, Newmont.com, August 12. http://www.newmont.com/ our-voice-blog/2013/Mining-101-Understanding-the-Different-Phases-of-Mine-Operations/default.aspx?view=details&item=Mining-101-Understanding-the-Differ-ent-Phases-of-Mine-Operations.

Nordhaus, William D., and Edward C. Kokkelenberg, eds. 1999. *Nature's Numbers: Expanding the National Economic Accounts to Include the Environment*. Washington, DC: National Academy Press.

Noreng, Oystein. 2005. "Norway: Economic Diversification and the Petroleum Industry." In *The Gulf Oil and Gas Sector: Potential and Constraints*. Abu Dhabi, UAE: The Emirates Center for Strategic Studies and Research.

OECD (Organisation for Economic Co-operation and Development). 2010. "The Economic Impact of Export Restrictions on Raw Materials." http://dx.doi. org/10.1787/9789264096448-en.

———. 2014. "Volume Measurement of Stocks of Natural Resources." COM/ENV/ STD(2012)2 JT03365606, background note, OECD, Paris, France.

Openoil. "Oil Contracts: How to Read and Understand Them." http://openoil.net/under-standing-oil-contracts/.

Ortega Girones, Enrique, Alexandra Pugachevsky, and Gotthard Walser. 2009. "Mineral Rights Cadastre." World Bank Extractive Industries for Development Series #4, World Bank, Washington, DC. http://siteresources.worldbank.org/INTOGMC/Resources/ Mining_Cadastre_Revised.pdf.

Ossowski, Rolando, and Håvard Halland. Forthcoming, 2015. "Fiscal Management in Resource-Rich Countries." Vol. 2 of *Essentials for Economists, Public Finance Professionals, and Policy Makers*. World Bank Studies Series. Washington, DC: World Bank.

Otto, James, Frank Stermole, Fred Cawood, Pietro Guj, John Stermole, Michael Doggett, Craig Andrews, and John Tilton. 2006. "Mining Royalties: A Global Study of Their

Impact on Investors, Government, and Civil Society." Report No. 37258, World Bank, Washington, DC. http://documents.worldbank.org/curated/en/2006/06/7045893/mining-royalties-global-study-impact-investors-government-civil-society-vol-1-2.

Parthemore, Christine. 2011. *Elements of Security Mitigating the Risks of U.S. Dependence on Critical Minerals.* Washington, DC: Center for a New American Security. http://www.cnas.org/files/documents/publications/CNAS_Minerals_Parthemore.pdf.

Peck, Philip, and Knut Sinding. 2009. "Financial Assurance and Mine Closure: Stakeholder Expectations and Effects for Operating Decisions." *Resources Policy* 34: 227–333.

Proserv Offshore. 2010. *Decommissioning Cost Update for Removing Pacific OCS Region Offshore Oil and Gas Facilities.* Final Report, Volume 1, Proserv Offshore, Houston, TX.

Provincial Government of Western Cape. 2005. "Guideline for Environmental Management Plans." Cape Town, South Africa. http://www.westerncape.gov.za/Text/2005/7/deadp_emp_guideline_june05_5.pdf.

Queensland Government. Undated. Guideline (a), "Preparing an Environmental Management Plan (Exploration Permit or Mineral Development License) for a Level 1 Mining Project."

Rajaram, Anand. 2012. "Improving Public Investment Efficiency." Presentation to the IMF conference "Management of Natural Resources in Sub-Saharan Africa," Kinshasa, March 21–22.

Rajaram, Anand, Tuan Minh Le, Nataliya Biletska, and Jim Brumby. 2010. "A Diagnostic Framework for Assessing Public Investment Management." Policy Research Paper 5397, World Bank, Washington, DC. http://elibrary.worldbank.org/doi/pdf/10.1596/1813-9450-5397.

Rajaram, Anand, Tuan Minh Le, Kai Kaiser, Jay-Hyung Kim, and Jonas Frank, eds. 2014. *The Power of Public Investment Management: Transforming Resources into Assets for Growth.* Washington, DC: World Bank.

Reedman, A. J., R. Callow, D. P. Piper, and D. G. Bate. 2008. "The Value of Geoscience Information in Less Developed Countries." Research Report CR/02/08, British Geological Survey, Nottingham, UK.

Rendu, Jean-Michel. 2014. *An Introduction to Cutoff Grade Estimation.* 2nd ed. Englewood, CO: Society for Mining Metallurgy and Exploration.

ResourceContracts.org. 2014. "Mining Contracts: How to Read and Understand Them." http://www.resourcecontracts.org.

Ricardo, David. 1821. "On the Principles of Political Economy and Taxation. Library of Economics and Liberty." Retrieved November 29, 2014. http://www.econlib.org/library/Ricardo/ricP1a.html.

Rio Tinto. 2013. *Rio Tinto 2013 Chartbook.* London, England: Rio Tinto Plc. http://www.riotinto.com.

Robinson, James A., Ragnar Torvik, and Thierry Verdier. 2006. "Political Foundations of the Resource Curse." *Journal of Development Economics* 79 (2): 447–68.

Rosenblum, Peter, and Susan Maples. 2009. *Contracts Confidential: Ending Secret Deals in the Extractives Industries.* New York: Revenue Watch Institute.

Ross, Kaiser, and Nimah Mazaheri. 2011. "The 'Resource Curse' in MENA? Political Transitions, Resource Wealth, Economic Shocks, and Conflict Risk." World Bank, Washington DC.

Sachs, Jeffrey, and Andrew Warner. 1995. "Natural Resource Abundance and Economic Growth." NBER Working Paper 5398, National Bureau of Economic Research, Cambridge, MA.

Sassoon, Meredith. 2009. "Guidelines for the Implementation of Financial Surety for Mine Closure." World Bank, Washington, DC. http://siteresources.worldbank.org/ INTOGMC/Resources/7_eifd_financial_surety.pdf.

SEADI (Support for Economic Analysis Development in Indonesia). 2013. "The Economic Effects of Indonesia's Mineral-Processing Requirements for Export." USAID, Washington DC. http://pdf.usaid.gov/pdf_docs/pbaaa139.pdf.

Smith, E., and P. Rosenblum. 2011. *Enforcing the Rules: Government and Citizen Oversight of Mining.* New York: Revenue Watch Institute (now the Natural Resource Governance Institute). http://www.revenuewatch.org/sites/default/files/RWI_Enforcing_Rules_ full.pdf.

SPE (Society of Petroleum Engineers), AAPG (American Association of Petroleum Geologists), WPC (World Petroleum Council), SPEE (Society of Petroleum Evaluation Engineers), and SEG (Society of Exploration Geophysicists). 2011. "Guidelines for Application of the Petroleum Resources Management System." http://www.spe.org/ industry/docs/PRMS_Guidelines_Nov2011.pdf.

Stanley, Michael, and Adriana Eftimie. 2005. *Government Support for Sustainability of Extractive Industries.* Washington, DC: World Bank.

Stanley, Michael, and Ekaterina Mikhaylova. 2011. "Mineral Resource Tenders and Mining Infrastructure Projects: Guiding Principles." Working Paper 65503, World Bank, Washington, DC. http://documents.worldbank.org/curated/en/2011/09/15476393/ mineral-resource-tenders-mining-infrastructure-projects-guiding-principles.

Statistics Canada. 2006. "Concepts, Sources, and Methods of the Canadian System of Environmental and Resource Accounts." Catalogue No. 16-5-5-GIE. http://www. statcan.gc.ca/pub/16-505-g/16-505-g1997001-eng.pdf.

Tilton, John. 1985. "The Metals." In *Economics of the Mineral Industries,* 384–415. 4th ed. New York: AIME.

Tordo, Silvana, David Johnston, and Daniel Johnston. 2010. "Petroleum Exploration and Production Rights: Allocation Strategies and Design Issues." Working Paper 179, World Bank, Washington, DC. http://elibrary.worldbank.org/doi/pdf/10.1596/978-0-8213-8167-0.

Tordo, Silvana, Michael Manzano Warner, and Yahya Osmel Anouti. 2013. " Local Content Policies in the Oil and Gas Sector." http://elibrary.worldbank.org/doi/pdf/10.1596/978-0-8213-9931-6.

Torvik, Ragnar. 2001. "Learning by Doing and the Dutch Disease." *European Economic Review* 45: 285–306. http://www.sv.ntnu.no/iso/Ragnar.Torvik/science.pdf.

———. 2002. "Natural Resources, Rent Seeking, and Welfare." *Journal of Development Economics* 67: 455–70.

UN (United Nations). 2010. "United Nations Framework Classification for Fossil Energy and Mineral Reserves and Resources 2009," 39. ECE Energy Series, United Nations Economic Commission for Europe, Geneva, Switzerland. http://www.unece.org/ energy/se/unfc_2009.html.

UN, EU (European Union), FAO (Food and Agriculture Organization of the United Nations), IMF, OECD, and World Bank. 2014. "System of Environmental-Economic

Accounting 2012—Central Framework." United Nations, New York. http://unstats.
un.org/unsd/envaccounting/seeaRev/SEEA_CF_Final_en.pdf.

UNCTAD (United Nations Conference on Trade and Development). 2013. " Time Series
on Inward and Outward Foreign Direct Investment Flows, Annual, 1970–2012." Data
compiled by the *Financial Times*, August 19, "Offshore Centres Race to Seal Africa
Investment Tax Deals." http://www.ft.com/intl/cms/s/0/64368e44-08c8-11e3-ad07-
00144feabdc0.html.

van der Ploeg, Frederick. 2011. "Natural Resources: Curse or Blessing?" *Journal of
Economic Literature* 492: 366–420.

van der Ploeg, Frederick, and Steven Poelhekke. 2010. "The Pungent Smell of 'Red
Herrings': Subsoil Assets, Rents, Volatility, and the Resource Curse." *Journal of
Environmental Economics and Management* 60: 44–55.

Westgate, Leon. 2014. "2014 LME Week Annual Report." Standard Bank Group, London,
England.

Wood Mackenzie. 2013. "C1 Composite Cost Curve of Copper Mine Production (2013)."
Graphical representation.

World Bank. 2010a. *Mining Community Development Agreements—Practical Experiences
and Field Studies*. Washington, DC: World Bank. http://www.sdsg.org/wp-content/
uploads/2011/06/CDA-Report-FINAL.pdf.

———. 2010b. *Mining Foundations, Trusts and Funds: A Sourcebook*. June 2010.
Washington, DC: World Bank. http://siteresources.worldbank.org/EXTOGMC/
Resources/Sourcebook_Full_Report.pdf.

———. 2010c. "Towards Sustainable Decommissioning and Closure of Oil Fields and
Mines: A Toolkit to Assist Government Agencies." http://siteresources.worldbank.org/
EXTOGMC/Resources/336929-1258667423902/decommission_toolkit3_full.pdf.

———. 2011a. "Overview of State Ownership in the Global Minerals Industry." Extractive
Industries for Development Series #20, World Bank, Washington, DC. http://sitere-
sources.worldbank.org/INTOGMC/Resources/GlobalMiningIndustry-Overview.pdf.

———. 2011b. *Sharing Mining Benefits in Developing Countries. The Experience with
Foundations, Trusts, and Funds*. Extractive Industries for Development Series No. 21.
Washington, DC: World Bank. http://www-wds.worldbank.org/external/default/
WDSContentServer/WDSP/IB/2011/06/14/000333037_20110614052552/
Rendered/PDF/624980NWP0P1160ns00trusts0and0funds.pdf.

World Bank Institute. "Extractives Industries Contract Monitoring Roadmap." World Bank,
Washington, DC.

Zientek, Michael L., James D. Bliss, David W. Broughton, Michael Christie, Paul D.
Denning, Timothy S. Hayes, Murray W. Hitzman, John D. Horton, Susan Frost-Killian,
Douglas J. Jack, Sharad Master, Heather L. Parks, Cliff D. Taylor, Anna B. Wilson, Niki
E. Wintzer, and Jon Woodhead. 2014. "Sediment-Hosted Stratabound Copper
Assessment of the Neoproterozoic Roan Group, Central African Copperbelt, Katanga
Basin, Democratic Republic of the Congo and Zambia." Scientific Investigations
Report 2010–5090–T, Prepared by the U.S. Geological Survey, Reston, Virginia.
http://dx.doi.org/10.3133/sir20105090T.

green
press
INITIATIVE

www.ingramcontent.com/pod-product-compliance
Lightning Source LLC
Chambersburg PA
CBHW082357270326
41935CB00013B/1663